MAN, NATURE AND TECHNOLOGY

Also by Erik Baark

INDIA–CHINA COMPARATIVE RESEARCH (*editor with Jon Sigurdson*)
*TECHNOLOGICAL DEVELOPMENT IN CHINA, INDIA AND JAPAN (*editor with Andrew Jamison*)

Also by Uno Svedin

RESOURCES, SOCIETY AND THE FUTURE (*co-author*)
METHODS IN FUTURES STUDIES – PROBLEMS AND APPLICATIONS (*with B. Schwarz and B. Wittrock*)
NATURE AS A SYMBOL (*in Swedish*)
EARTH – MAN – HEAVEN (*in Swedish*)

*Also published by Macmillan

Man, Nature and Technology

Essays on the Role of Ideological Perceptions

Edited by

Erik Baark

Research Fellow
Research Policy Institute
University of Lund

and

Uno Svedin

Executive Secretary
Committee for Natural Resources Research
Swedish Council for Planning and Coordination of Research

Foreword by Torsten Hägerstrand
Chairman of SALFO, 1974–85

MACMILLAN
PRESS

First published 1988

Published by
THE MACMILLAN PRESS LTD
Houndmills, Basingstoke, Hampshire RG21 2XS
and London
Companies and representatives
throughout the world

Typeset by Wessex Typesetters
(Division of The Eastern Press Ltd)
Frome, Somerset

British Library Cataloguing in Publication Data
Man, nature and technology: essays on the
role of ideological perceptions.
1. Technology—Philosophy
I. Baark, Erik   II. Svedin, Uno
601   T14
ISBN 978-1-349-09089-1   ISBN 978-1-349-09087-7 (eBook)
DOI 10.1007/978-1-349-09087-7

Contents

Foreword

If human culture from its very beginning had been uniformly the same everywhere, it would have been extremely hard to discern. We discover culture when we experience world-views and ways of life which differ between continents and regions. The individuality of a culture is the outcome of an historical process, rooted in the characteristics of the physical environment, coloured by a unique position in the global flow of ideas, and protected by a certain amount of isolation. The permeable barrier around a culture is of particular significance. A completely isolated group lacks the stimulus of communication. But without some protection the individuality is gradually eroded. In today's world of instant information and global circulation of goods, language is perhaps the only remaining membrane which simultaneously separates and connects.

The sciences and the science-based technologies have developed their own set of cultural identities, each of which uses a language of its own. In this respect we find a striking similarity with cultural development in general. But there is also a fundamental difference. Our everyday taken-for-granted culture is parochial. It reflects the experiences of people in their historically established environments. The scientific cultures, on the other hand, claim to be global in nature. They divide the world into uniform strata of standardised concepts.

Today's world is permeated by a strong interaction between the two forms of culture. Each form affects the other, but neither is likely to become totally dominant. There is far too much in human life which cannot – or should not – become subsumed into the rules of scientific cultures. And, in fact, science and technology are by no means so independent of historically established cultural circumstances as their spokesmen like to believe.

Since the use of natural resources is a human necessity, every historical culture has developed its own ideas of nature. Clearly, the spread of a scientific mode of thinking and acting will profoundly influence these ideas and, to a certain extent, be affected by them. One of the problematic outomes of the shift towards globalisation of concepts and practical standards is that people will have less and less opportunity to see for themselves the consequences in nature of their actions. In former days when redistribution of matter was less extensive, nature could, so to speak, send back messages to the inhabitants of an area and thereby modify their outlook and behaviour. The emerging situation makes the loops of cause and effect increasingly indistinct and incomprehensible. Systematic monitoring, of course, can help to a certain extent. But an enduring problem will be that our institutions for decision-making and control are inherited from the local cultures and not easily adapted to interrelations of global dimensions.

The considerations just exemplified – many of which are particularly relevant for a small country – caused the Committee for Future Oriented Research (SALFO) at the Swedish Council for Planning and Coordination of Research (FRN) to launch a programme composed of a series of trans-disciplinary studies of the complex relations that exist between culture, nature and technology. The approach has been historical as well as comparative. In the course of this work the Swedish researchers have had the advantage of inspiration from many colleagues outside Sweden. In order to make some of the contributions of this group more generally available the Committee asked Dr Uno Svedin (co-ordinator of the project 'Natural resources in a cultural perspective') and Dr Erik Baark (co-ordinator of the project 'Technology and culture') to promote a synthesis by assembling and commenting upon a collection of essays dealing with the role of ideology in the developments within the culture–nature–technology triangle. Thanks are due to the authors of the essays for their generous collaboration and to the two project leaders and editors for their dedicated work over several years.

Lund

TORSTEN HÄGERSTRAND
Chairman of SALFO, 1974–85

Notes on the Contributors

Erik Baark is Research Fellow at the Research Policy Institute, Unviersity of Lund, Sweden. His main interest has been research on the conditioning of technological development in non-Western countries, in particular China, Japan and India. He believes that research on these processes of accommodating technological change to the specific social and cultural characteristics obtaining is important for the study of technology in relation to human history. Taking a special interest in the development and use of information technologies, he believes that differences in patterns of exchange and processing of information arise from, and at the same time condition, the development of modern society. His publications include *India–China Comparative Research* (1981) which he edited with Jon Sigurdson, and *Technological Development in China, India and Japan* (1986) which he edited with Andrew Jamison.

Robert W. Kates is a geographer and is currently University Professor and Director of the Alan Shawn Feinstein World Hunger Program at Brown University, Providence, Rhode Island, USA. He is the author of papers, books and monographs on geophysical and technological hazards; landscape perception; water resources and rural development, particularly in Africa; he has also helped develop the methodologies of environmental perception, risk assessment, and climate impact assessment. The paper presented in this volume derives from his long-term interest in theories of human environment. This theoretical work is complemented by a series of current empirical studies dealing with long-term human environmental experience on scales of centuries and millenia. Specifically these include studies of fluctuations in human population, the last three hundred years of human-induced earth transformation, and the identification of potential limits, if any, to the sustainable development of the biosphere over the next century. His publications include many papers related to the human environment, and the joint-editorship of *Perilous Progress: Technology as Hazard* (1985).

David Lowenthal is Professor of the Department of Geography, University College, London, United Kingdom. He has long had active links with crusades for conserving both nature and antiquities, our physical and our cultural heritage. Since the 1950s he has been involved with such organisations as the Conservation Foundation, Resources for the Future, the International Council on Monuments and Sites, Save Britain's Heritage, and the Caribbean Conservation Association. Yet his major concern is less with saving things than with trying to understand the history of impulses to conserve, the background of historical and ecological understanding, and the motivations that fuel these crusades. These interests are embodied in his publications, which include the biography of the pioneer American conservationist Marsh, *George Perkins Marsh: Versatile Vermonter* (1958), the edition of Marsh's 1864 classic, *Man and Nature* (1965), and his two recent books on historical heritage, *Our Past Before Us: Why Do We Save It?* (1981) and *The Past Is a Foreign Country* (1985).

Francisco R. Sagasti is Director of the Grupo de Analisis para el Desarollo (GRADE), Lima, Peru. His initial professional and intellectual interests centred upon the use of management science and operations research in development planning and management. During the last fifteen years he has worked in the field of science and technology in developing countries, examining issues such as the historical evolution of science in Latin America, the science and technology policy design and implementation process in developing countries, and international collaboration in science and technology. He has been one of the key researchers in the 'Science and Technology Policy Indicators' project which was initiated by the Science Policy Research Unit, Sussex University, United Kingdom. He was also deeply involved in the debates which were associated with the United Nations Conference on Science and Technology for Development, convened in Vienna in 1979. His more recent work has focussed on the problem of identifying and designing development options and strategies, seeking to develop new approaches and methods for long-term planning in developing countries. Simultaneously, his interests are shifting towards providing an integrated perspective of the role of modern science and technology in the process of development, a concern which is reflected in his contribution to this volume. He has published fourteen books and monographs and more than one hundred papers, most of which relate to science and technology for development.

Uno Svedin is Executive Secretary of the Committee for Research on Natural Resources at the Swedish Council for Planning and Coordination of Research. He was trained as a physicist and holds a doctorate in physics at the University of Stockholm. He later specialised in technology assessment and natural resources issues and was a member of the project group on 'Resources and Raw Materials' at the Secretariat for Futures Studies during the 1970s. The final report 'Resources, Society and the Future' was published in 1980 in its English version by Pergamon. Svedin was responsible for the preparation of the Swedish National Report to the 1979 UN Conference on Science and Technology for Development (UNCSTD). His interests include methods of futures studies (he has published *Methods in Futures Studies – Problems and Applications*, 1983, together with B. Schwarz and B. Wittrock), technology assessment and environmental perception. Books (in Swedish) include *Nature as a Symbol*, 1983, and *Earth – Man – Heaven*, a few scholars' views of Nature as presented in the major cultures of the world.

Michael Thompson is Principal Research Fellow, Institute for Management Research and Development, University of Warwick, and Technology Policy Unit, University of Aston, United Kingdom. He is an anthropologist by training and applied systems analyst (or policy analyst) by occupation. Over the last ten years or so he has been engaged in a programme of interdisciplinary, international and often collaborative research, his feeling being that if anthropology is indeed the queen of social sciences, it should have something valuable to contribute to our understanding of technology, information, organisation, management and leadership. It should contribute to the whole business of policy formation and policy viability, and to the economics and aesthetics of social life ancient *and* modern (and post-modern). He has worked on substantive issues in a wide range of policy areas: environmental policy, social policy, energy policy, technology policy and (most recently of all) industrial and business policy. At the same time, he has worked to harness all this applied research to the parallel development of the broad theoretical framework which, in his view, anthropology alone can provide. As a leader of the Core Concepts Project at the International Institute for Applied Systems Analysis (IIASA), he has worked to develop and communicate this framework, components of which are also presented in his contribution to this volume. His publications include *Rubbish Theory: The Creation and Destruction of Value* (1979) and many papers and articles exploring the cultural dimensions of risk, policy-making, and so on.

Donald Worster is Jack E. Meyerhoff Professor of American Environmental Studies, Brandeis University, USA. He has primarily been interested in the new field of environmental and ecological history – that is, history viewed from an ecological perspective. For a long while historians have tended to treat human history as though it had nothing to do with the earth and its processes. Worster calls for a perspective which sees the fate of societies as dependent on such natural variables as climate, soil, vegetation, ecosystems, the carbon cycle and the like. People should realise too that what humans do to the earth can have serious adverse consequences, for the earth as well as for the people. This is an important theme in American social history, but today it has world-wide significance, as marginal lands around the globe are being exploited, often by market-oriented developers, and desertification has become an international problem. Worster's publications include the recent analysis of issues similar to those covered in the chapter presented in this volume, *Rivers of Empire: Water, Aridity, and the Growth of the American West* (1986). He has also published *Nature's Economy: A History of Ecological Ideas* (1985; orig. publ. 1977) and *Dust Bowl: The Southern Plains in the 1930s* (1979).

Brian Wynne is Reader in Science Studies, and Director of the Centre for Science Studies and Science Policy, at the University of Lancaster, United Kingdom. Originally a materials scientist, he has taught and researched on the social aspects of science and technology since 1974. In 1977 he took part as a lay advocate in the Windscale nuclear plant Public Inquiry, and later wrote a book about it, *Rationality and Ritual: the Windscale Inquiry and Nuclear Decisions in Britain* (1982), which was described by Alvin Weinberg, elder statesman of nuclear energy and United States science policy, as brilliant, but marred by an anti-nuclear tone! In 1983–4 Brian Wynne led a research group on Institutional Settings and Environmental Policies at the International Institute for Applied Systems Analysis (IIASA), Austria, where he worked closely with Michael Thompson. Several publications have appeared from this project, including a special issue of *Policy Sciences* (December 1984) on technical and institutional problems in global energy modelling. Apart from the above-mentioned book on the

Windscale inquiry, his publications include the report from IIASA, *Hazardous Management of Wastes: Implementation and the Dialectics of Credibility* (1987).

1 Introduction

Erik Baark and Uno Svedin

I THE THEME OF THE BOOK

Most attempts to comprehend man's relationship to nature and technology have tended to be biased in the direction of either the environmental sciences or the engineering sciences. Many scholars are beginning to appreciate, however, that such partial approaches fail to capture essential aspects of the increasingly complex problems generated as man extends his sphere of influence into that of nature. A more comprehensive perspective is required in order to deal with this crisis of understanding and managing the linkages between nature, culture, and technology. Such a perspective is provided, we would argue, by seeing man's relationship to nature and technology as partly conditioned by ideological perceptions.

Nature and technology constitute essential components of human identity. In some way every human being defines his or her social and cultural roles in relation to these components. The human environment plays a crucial role in the main doctrines of world religions. Islam identifies nature with signs of God, and similar ideas are found in most other major religions. A number of religious and philosophical schools of thought in the East Asian countries, such as the Chinese philosophy of Taoism, are characterised by a conscious and active relationship with nature.[1] Modern secularism and exploitation of natural resources seem far removed from such attitudes, however, and in the eyes of many observers widespread devastation of the human environment is a likely outcome of this exploitation.

Technology is also integrated into the lives of human beings; technologies of every conceivable kind are used everywhere by human beings in order to provide food, shelter, transportation, and all other basic material appurtenances of life.[2] Technological change has become vital for modern economic development and, simultaneously, a growing concern with the role of technology in society has emerged. This concern has become apparent both in the search for public policies concerning technological development, and in the formation of social movements aiming at the control of technology in society.

The theme of this book is man's perceptions of nature and

1

technology – in particular, perceptions as contingent upon cultural attitudes and ideologies. Ideology is usually not a very precisely defined concept. We prefer to use the concept in the broad sense of a set of perspectives on human life or culture.[3] One should not, however, consider ideology as something homogeneous, that is, a set of logically coherent ideas. On the contrary, ideologies do tend to become constituted more as a kind of collage – a variety of ideas, sometimes self-contradictory, put together in a rough totality.

Ideologies often provide frameworks for possible modes of action. At the same time, however, much of the actual practice *vis-à-vis* nature and technology is not expressed in explicit terms – it simply consists of things that are done. Similarly, ideologies are frequently loaded with emotions and constructed primarily on the basis of a partial, rather than a complete, understanding of the questions addressed.

Ideologies are therefore important because they allow man to interpret his or her environment and to guard against the uncertainties of actions in the present and the future. This is especially clear in the case of the collective aspects of ideology. It is possible to speak about the ideology of an individual, but usually this implies that this person's perspectives are similar to some already existing pattern of ideological views. Despite the fact that most ideologies can only be rather loosely defined as a set of views, they will often have a direct impact on various forms of collective action, for example in a parliament, or in relation to public or private decision-making regarding the choice of technology.

The emerging field of biotechnology provides a fairly clear example of the role of ideologies in relation to nature and technology. There is growing concern for the preservation of nature's genetic diversity as embodied in the large variety of species existing on Earth. The technologies of genetic engineering promise to provide man with a tool to manipulate the composition of life forms. However, the artificially produced resources of future biotechnology can be seen as dependent on the wide reservoir of natural riches developed during the process of evolution. To what extent one perceives this reservoir as being freely available for exploitation by means of very advanced biotechnology is, we would argue, largely a question of ideological points of view.

Similarly, the discussion of 'limits to growth' has raised the question of whether the seemingly large domain of nature on this planet is large enough for an ever-growing population. The formulation of the problem as well as the solutions offered by those who advocate that the capability of technological growth can fill the gap of diminishing resources have strong ideological overtones. Resources as such are subject to definitions which draw heavily on cultural or ideological frameworks: can they be defined independently of technological levels or social needs? Many of us would argue that this is not possible, while others would claim that basic resources are fixed, referring only to those elements with which nature, 'untouched by man', will provide us.

II A BRIEF OVERVIEW OF THE BOOK

The first essay by Robert W. Kates serves to introduce many of the key concepts and issues taken up in connection with the interaction of nature, society, and technology. This concerns resource adequacy, environmental degradation in relation to development, and whether technological change threatens to overwhelm us through its massive force and through the complexity involved. To integrate and explain the many diverse phenomena, and thus go beyond the surface manifestations in order to reach the underlying processes, is a central role of theory. Kates's essay accordingly surveys and classifies major theories dealing with these processes. They include one-dimensional theories of causality, dichotomous or partial theories, and finally a series of interactive models. The author concludes that a distinctive theory of the human environment has yet to be enunciated, and suggests some directions for the development of such theory.

The second essay, by Francisco R. Sagasti, indicates that major advances in science and technology at the world level during the last thirty years make it necessary to reinterpret the concept of 'development' and to offer explanatory schemes which explicitly incorporate the process of knowledge generation. This essay accordingly constitutes a preliminary attempt to integrate issues related to science and technology with the conceptualisation of the development process. The starting-point is a critique of the general model offered by George Basalla to explain the diffusion of Western science. This is followed by a brief description of the components of an alternative conceptual framework, and the outline for an explanatory scheme that would link the various components of the alternative model in an organic fashion.

In the essay written by Michael Thompson the main theme of the book is explored on the basis of the cultural hypothesis. This notion brings together insights from two areas of inquiry. One of these is the sociology of perception. Here viable ideas of nature are traced back to the different sets of moral justifications necessary for the maintenance of social organisation. The other is a perspective centring on the ecology of natural resources. Various types of ideas of nature, and social groups associated with these, are categorised in a conceptual scheme (grid–group matrix), which in turn is employed in an analysis of the formation of policy. The perceptions of risk and resources in the energy debate illustrate the impact of such ideas of nature.

In the following essay Brian Wynne argues that technology embodies a differentiated set of cultures. Thus it will imply associated attitudes, images and belief systems which legitimise the social relations of technology. People react to technology by creating defensive images, for example technological animism which projects the character of the living world into technological systems. In this way some technologies are regarded as being beyond control by man. Recent experiences of ecological and technological disasters are analysed, and their effects on psychological and social roles described.

The essay by Donald Worster attempts to trace the origins of images of nature and technology to power relations. The author uses a case study of a modern desert society, the American Southwest, to show the links between ecology and power. Democracy, as it evolved out of Western market-place cultures, carried along with it an ideology of nature domination, maximum exploitation of resources, and instrumentalism. That ideology, the author argues, has built into it certain implications for power accumulation. Whatever group can bring about the domination of nature most effectively earns, by the terms of the ideology, a claim to considerable power. The results, as seen in American irrigation communities, when evaluated by other criteria, can be profoundly anti-democratic.

David Lowenthal argues that campaigns to save natural resources and artifactual heritage show significant parallels in their origins, agents, development, and motivations. After seeking to account for these similarities, the author examines how and why concerns manifested for natural and cultural relics also differ in character, degree, and effectiveness. Nature conservation arguments tend to be based on utilitarian principles of an economic or ecological kind, whereas the preservation of antiquity is more commonly justified on cultural and spiritual grounds. Shifting balances between principles of stewardship and self-interest, and growing awareness of the nature and consequences of human impact affect reactions to the threatened loss of both kinds of patrimony.

The theme of the book, and a number of dimensions which this theme represents in relation to nature, culture and technology, are taken up in the epilogue. Issues illustrating the theme and its implications for research are discussed under three headings. For example, we raise issues concerning idealistic or materialistic approaches and their connection with causal linkages, the role of power relations and risk perceptions, and finally the role of Utopian visions in providing guiding images for the future.

Notes

1. We see no reason to commit ourselves or the other contributors to a fixed definition of the concept of nature. In order to give a rough idea of our own position, we would like to cite the six aspects of the term 'nature' identified in the project description *Natural resources in a cultural perspective* (Stockholm: Committee for Future Oriented Research, 1981), p. 57: (1) Totality, reality; (2) Part of the environment which is unaffected by man; (3) The non-artifical; (4) Essence; (5) Harmony, lack of disturbance and strain; (6) The expected, explicable or evident.

2. Technology has also been provided with numerous definitions. A concise formulation is that technology 'is simply a body of knowledge about techniques' in C. Freeman, *The Economics of Industrial Innovation* (Harmondsworth: Penguin, 1974) p. 18. A more detailed definition, which we feel is fairly informative, is the following: 'Technology denotes the broad area of purposeful application of the contents of the physical, life, and behavioural sciences. It comprises the entire notion of technics as well as the medical, agricultural, management and other fields with their total hardware and software contents' in E. Jantsch, *Technological Forecasting in Perspective* (Paris: OECD, 1967) p. 15. A more comprehensive discussion of technological development is presented in 'The Technology and Culture problematique' in E. Baark and A. Jamison (eds), *Technological Development in China, India and Japan* (London: Macmillan, 1986).

3. A reference work such as *Webster's New Collegiate Dictionary* defines 'ideology' as '1: visionary theorizing; 2(a): a systematic body of concepts especially about human life or culture, (b): a manner or the content of thinking characteristic of an individual, group, or culture, (c): the integrated assertions, theories and aims that constitute a sociopolitical program.' The concept has also changed from its earliest meaning similar to 'science of ideas' to a pejorative sense implying a sense of abstract, impractical or fanatical theory. This pejorative sense has persisted until

today, when ideology is often contrasted with 'science'. See the discussion by Raymond Williams in *Keywords: A Vocabulary of Culture and Society* (New York: Oxford University Press, 1976).

2 Theories of Nature, Society and Technology*

Robert W. Kates

This quest for theory arises from the practice of environmental management. It began in the early 1970s, when Torsten Hägerstrand and I attempted to initiate a working group on 'theories and concepts of man and environment', a 'modest experiment in mutual encouragement' that essentially failed. However, the major reason for undertaking the initiative is still valid, albeit perhaps with lessened urgency.

In 1973 the just completed United Nations Conference on the Human Environment (known widely as the Stockholm Conference) provided a forum for the industrialised countries' concerns with national and global issues of environment. The range of global issues so defined appear in a series of annual 'State of the Environment' reports and these indicate the variety of concerns that arose under the rubric of environment (Holdgate, Kassas and White, 1982). See Table 2.1.

The annual report of the Executive Director of the United Nations Environment Programme is one distillation of the problems of the human environment. Another, perhaps more relevant to everyday experience, is media reports. For example, in a recent two-week period between 65 and 75 different technological hazards were discussed in the popular electronic and print media of my small city.

It was apparent to Torsten and myself, even in 1973, that people were being overwhelmed by candidate worries. What to worry about had become a central question of our time. In a sense, many of us are professional worriers, appointed or self-selected to help society to identify, measure, and weigh worries, to anticipate what to take seriously and what to ignore, what to act upon and what to study more. A quasi-discipline or profession has arisen, with training, literature, research and societies devoted to the assessment of environmental impact, technology, social impact and risk. The fields

* I am grateful to my fellow conference participants and to B. L. Turner III for many helpful comments, but most of all to a generation of graduate students in geography at Clark University who have shared the search for theory with me.

Table 2.1   Topics treated in annual State of the Environment reports

Subject area	Topic	Year
The atmosphere	Climatic changes and their causes	1974*, 1976, 1980
	Possible effects of ozone depletion	1977
The marine environment	Oceans	1975*
Freshwater enrivonment	Water resources and quality	1974*, 1976
	Ground Water	1981
Land environment	Land resources	1974*
	Raw materials	1975*
	Firewood	1977
Food and agriculture	Food shortages, hunger, and losses of agricultural land	1974*, 1977, 1976
	Use of agricultural and agro-industrial residues	1978
	Resistance to pesticides	1979
Environment and health	Toxic substances and effects	1974*, 1976*
	Heavy metals and health	1980
	Cancer	1977
	Malaria	1978
	Schistosomiasis	1979
	Biological effects of ozone depletion	1977
	Chemicals in Food Chain	1981
Energy	Energy conservation	1975*, 1978
	Firewood	1977
	—possible effects of ozone depletion	1977
	—chemicals and the environment	1978
	Noise pollution	1979
Man and environment	Human stress and societal tension	1974*
	Outer limits	1975*
	Population	1975*, 1976*
	Tourism and environment	1979
	Transport and the environment	1980
	Environmental effects of military activity	1980
	The child and the environment	1980
Environmental management achievements	The approach to management	1974*, 1976
	Protection and improvement of the environment	1977
	Legal and institutional arrangement	1976*
	Environmental economics	1981

* Indicates brief treatment in early reports.
Source: Holdgate *et al.* (1982).

have grown rapidly. For example, by 1983 the field of comparative risk analysis (Kates and Kasperson, 1983), which can be dated to a seminal paper published in 1969 (Starr, 1969) had produced in the English language alone, 54 book-length volumes, and bibliography listing over 1000 items (Covello and Abernathy, 1983).

Despite this outpouring of scientific effort, professional help concerning what to worry about (as opposed to personal strategies) is often confused or contradictory. This arises, in part, from the internal sociology of science and its interaction with society, partly from the complex nature of the issues themselves, and partly from fundamental lack of understanding. For example, it is fundamentally simpler to design a safe bridge than to design a safe chemical, given the difference in knowledge of principles of structural analysis and principles of carcinogenesis. There is no consensual theory of carcinogenesis or atherosclerosis and there is certainly no consensual theory of broader issues, for example, the fragility or robustness of nature, the determinants of population growth and decline, the relationship between resources and societal well-being, or between envirionment and development. Thus, while much can be done to inform societal understanding of candidate worries by empirical findings or professional efforts at analysis, long-term progress depends on theoretical understanding.

As with carcinogenesis, the absence of a consensual theory does not denote a shortage of theory. Theories of the human environment abound today as they did in 1973 or, for that matter, in 1873 or even 73. However in 1973 we felt that most theories were 'transfer' theories, derived from either the natural or physical sciences (primarily ecology, engineering and physics) or from the social and behavioural sciences (primarily economics) and applied to the joint domain of man and environment. This bothered us and we felt, more by inspiration than erudition, that this important domain of human concern 'required' distinctive explanatory theory as a scientific aid to the social ordering of problems and issues.

So much for the origins of this quest for theory. The paper itself is in two parts. The first reviews existing theory within the framework of a simple, causal taxonomy. The second suggests directions for the development of distinctive, interactive theory and concludes with a set of theory-related empirical studies.

I NATURE, SOCIETY, TECHNOLOGY: A CAUSAL TAXONOMY

There is a long tradition of 'man–land' and later 'man–environment' studies in my discipline, geography (Burton, Kates and Kirkby, 1974). Initially, my conception of the domain of interest was much like that of Ward and Dubos in their eloquent background volume for the Stockholm Conference:

Man inhabits two worlds. One is the natural world of plants and animals, of soils and airs and waters which preceded him by billions of years and of which he is a part. The other is the world of social institutions and artefacts he builds for himself, using his tools and engines, his science and his dreams to fashion an environment obedient to human purpose and direction. (Ward and Dubos, 1972).

This latter world of social institution and artefact is close to the concept of culture as employed by anthropologists, and is broader than the use of culture implied by this volume (Williams, 1976). However, in reviewing theories of the human environment, it became clear that a significant number distinguished between 'social institutions' and 'artefacts'. Indeed, the relative autonomy of society and technology was a major point of contention. Thus, in what follows, I use as organising concepts nature, society and technology, rejecting an a priori subsumption of technology under social direction. One further modification of tradition is called for. The two worlds of Ward and Dubos appear to be René Dubos's worlds rather than Barbara Ward's; thus I employ the term 'human environment', for the world that surrounds both sexes.

Theories of the human environment can be arrayed in many ways. I choose to array them along a simple continuum of causality based on the major interactive factors of nature, society and technology. (See Table 2.2.) Under these categories, there are, firstly, one-dimensional theories of causality, in which one of the three interacting components is dominant. These include theories of supernatural control, biological determinism, environmental determinism, ecological balance, social dominance, and autonomous technology. More complex are dichotomous or partial theories which link nature and society, society and technology, and technology and nature, or single out some specific aspect of the interaction of nature, society, and technology. These partial theories cover an enormous array of the behavioural and social sciences. Thirdly there are a series of interactive models all of which encompass nature, society and technology, but do so at different levels of specificity and concept. These include conceptual models: triads and quadrads around such rubrics as place, work and folk or man, mind and land. Although such conceptual descriptions frequently use case studies, they lack specificity of detail and this is made up through accounting systems employing units of energy, materials, and information in various forms of input–output systems. Finally there are systems models, some conceptual, but most

Table 2.2   Theories of the human environment

ONE-DIMENSIONAL CAUSALITY
  Supernatural control
  Biological determinism
  Environmental determinism
  Ecological balance
  Social dominance
  Autonomous technology

PARTIAL THEORIES
  Nature and society
    Nature, nurture; resources/population
  Society and technology
    Productive forces, social relationships; stages and cycles of economic
      growth
  Technology and nature
    Human evolution; environmental crisis
  Nature, society and technology
    Factors of production, natural and technological hazards

INTERACTIVE THEORIES
  Conceptual triads and quadrads
    Place, work and folk; habitat, economy, and society; man, mind and
      land
    Population, organisation, environment and technology
    Knowledge, materials, energy; environment, subsistence, system,
      infrastructure, structure, superstructure
  Accounting systems
    Energy, materials, information
Systems models
    Biological systems, adaptive ecology, global and regional models

---

designed for computer simulation in which theory is often more
implicit than explicit.

One-dimensional causality

Supernatural control

The most widespread theory of the human environment is supernatural control. Invariably absent from texts on resource management or
environmental science, there is none the less widespread belief in the
creation, control, and direction of the human environment by
supernatural entities. These are enshrined in myths of creation
(Sproul, 1979), divided between earth gods and sky gods (Eliade,

1959), and ordained as dominion, imperatives and abominations (Douglas, 1966; Harris, 1966; White, 1967, Passmore, 1974, Diener and Robkin, 1978; Simoons, 1979) and even as science (Glacken, 1967, Kates, 1983a).

Biological determinism

Such theories root their explanations of the human environment in the primacy of biological relationships, primarily that of reproduction. In early forms they were linked to supernatural control – human fate was determined by God's grant of human dominion, or more selectively as one or another 'chosen' people, both ethnic and racial. With the advent of Darwin a biological process, natural selection, could now be posited as a driving force to explain differential success in human social life (Spencer, 1910). Rejecting social Darwinism and the inevitability of human competition, Kropotkin (1902) embraced its counterpart of co-operation through mutual aid, anticipating by more than half a century the search for a selective mechanism for altruism (Hamilton, 1963; Wilson, 1975) and the new biological determinism of the more extreme formulation of sociobiology (Caplan, 1978).

Environmental determinism

Paralleling biological determinism is theory that broadly holds that the natural environment, or some selected characteristic thereof, determines human social and physical characteristics. This is an ancient theory, embodied in Hippocrates's *Airs, Waters and Places* (Glacken, 1967). Climate and topography become favoured explanatory factors for cultural and historical events (Buckle, 1899; Semple, 1911; Huntington, 1965; Meggers, 1954). In more recent and less blatant determinism, the formulations are changed. Thus environment may determine livelihood, which in turn determines social organisation; or environment may only constrain probabilistically rather than determine; or subjective appraisals or perceptions of environment (in part culturally determined) may be the determinants of human behaviour (Burton, Kates and Kirkby, 1974).

Ecological balance

A blend of biological and environmental determinism is found in theories of ecological balance. Such theories view the economy of nature (Worster, 1977) as an interdependent and balanced economy, a balance struck between environmental potential and the community of plants and animals that form a specific ecosystem. Under natural conditions such ecosystems are stable. Through homeostatic mechanisms akin to the internal regulating mechanisms of human physiology (Cannon, 1932), ecosystems, when perturbed, will return to their equilibrium state. A corollary relates the diversity of ecosystems to their stability (Goodman, 1975). At its grandest scale, ecological balance is expressed as the Gaia hypothesis (Lovelock, 1979) which postulates an organic, evolving whole for all of this planet's living matter. The balance of nature is also a prescriptive theory for humankind, implying that nature indeed knows best (Commoner, 1971) and that human being disrupt such stability only at their peril. Thus, ecological balance can serve as a model for human society (Caldwell, 1969; Bookchin, 1971).

Social dominance

Theories of social dominance focus on the human capacity to tame or even create nature using a subservient technology. Beginning with the classic exposition of Marsh (1965; Lowenthal, 1958), expanded in the pathfinding symposium *Man's Role in Changing the Face of the Earth* (Thomas, 1956) and confirmed by recent studies of biogeochemical cycles (Holdgate, Kassas and White, 1982; Clark and Munn, 1986), there is widespread documentation of human domination of the natural environment. Within the theme of social domination are also corollary theories of domination: the domination of nature is but still another way of dominating men (Leiss, 1972) or women (Griffin, 1978).

Autonomous technology

Theories of social dominance presume a powerful but subservient technology, but theories of autonomous technology accord to *technique* (to use Ellul's broadened French conception) an independence of its own. The machine enters the garden (Marx, 1964) to take command (Giedion, 1948) of the technological society (Ellul, 1973). Technology creates its own autonomous imperative (Winner, 1978).

Partial theories

Nature and society

Dichotomous theories of nature and society have been developed within two main intellectual traditions, both fraught with controversy. The first revolves around 'nature and nurture' and each generation's restatement of the debate still evokes strident responses (Cronbach, 1975; Lewontin, Rose, and Kamin, 1984) despite the almost universal scientific recognition of genetic and environmental interaction, only the proportions of which are subject to investigation (Dubos, 1965).

The second major theoretical line is Malthusian (Petersen, 1979; James, 1979), although strictly speaking, the basic equation divides the denominator of population rather than social organisation by the resources of the numerator, resources that assume a level of technological proficiency. If any single theory can be said to dominate thought about the human environment, it is surely Malthusian theory and its widespread opposition (Simon, 1981). Central to neo-Malthusian debate is the widespread attention to the powerful metaphor of 'The Tragedy of the Commons' (Hardin, 1968).

Society and technology

Karl Marx is the outstanding theorist of society and technology. Bordering on technological determinism with the famous aphorism 'the handmill gives you the feudal lord' (Shaw, 1979), he nevertheless links productive forces (technology) to social relationships interactively, creating a compelling theory. The linkage between technology and society and, more specifically, economy, also preoccupies most other historically-oriented economists from diverse traditions for example, Joseph A. Schumpeter. In the face of the current global recession theorists as different as Forrester, Mandel and Rostow share a common interest in studying Kondratiev long waves, the multi-decadal fluctuations in the economy, attributed by most to fluctuation in the technological base of industrial economies (Freeman, Clark and Soete, 1982).

Technology and nature

Theories emphasising the links between technology and nature are used to explain the oldest and the newest human environments. Earliest human evolution is explained as interactions between technology and nature, perhaps because material culture and reconstructed environments dominate the older archaeological record.

At the other end of the temporal scale of the human environment are the 'exceptionalist' technological theories used to explain the environmental crisis of the recent decades (Commoner, 1971) and generating in turn the prescriptive alternative theories of appropriate technology (Schumacher, 1973; Lovins, 1976; Sachs, 1977).

Nature, society and technology

Finally there are a number of theories that are partial, not by their factorial emphasis but by their limited domain of application. Thus classical economic production theory combines nature, society and technology as land, labour and capital, but limits the explanation to the supply side of modern economies. Within this framework of classical theory, environmental problems are externalities of production and the solution to such problems is to internalise the costs of such externalities into the optimal mix of factors of production (Fisher, 1981). Another set of partial theories attempts to explain these aspects of the human environment related to natural hazard (Burton, Kates and White, 1978), employing limited concepts of nature, society and technology, a limitation for which they are severely criticised (Hewitt, 1983).

Interactive theories

Conceptual triads and quadrads

The most persuasive statements of interactive theories of nature, society and technology used words to express concepts and illustrate these with case studies. These are conceptual triads and quadrads built around variants of nature, society and technology: place, work and folk (LePlay, 1879), habitat, economy and society (Forde, 1963), man, mind and land (Firey, 1960), population, social organisation, environment and technology (Duncan, 1964), social behaviour, technology and resource opportunities (Butzer, 1982). Less obvious in the use of key words, but conceptually similar, is Boulding's (1981) use of materials, energy and information, Ellen's (1982) use of environment, subsistence, and system, and Harris's (1979) use of infrastructure, structure and superstructure.

Accounting systems

In contrast to conceptual theories, which are strong in concept and weak in explication are various accounting systems which embody the reverse: weak or implicit theory and strength of explication. They are primarily built around exchange values of energy (Odum, 1976, 1983), materials (Ayres, 1978), or monetary information (Leontief, 1977).

Systems models

These attempt to bridge the gap between triads and accounting systems. Models of systems are inevitably theories of systems although the theory is often implicit. There are at least three groups of models that seek to replicate interactions of nature, society and technology: biologically-based models, the most ambitious of which is Miller's *Living Systems* (1978), which hierarchically replicates and integrates systems ranging from the cell to the world order. Also biologically-based, but dealing with resource management, are the adaptive ecological models of Holling (1973) and his colleagues (Holling, 1978). Finally, there are the global and regional models of doom and hope (Meadows *et al.*, 1972. Cole *et al.*, 1973), the practice of which in its first decade was recently reviewed (Meadows, Richardson and Bruchmann, 1982).

The persistence of theory

Old theories, like old soldiers, never die, but are continuously rediscovered or redefined. Thus one-dimensional theories, while continually maligned for their simplistic quality and their flagrant misuse in the service of ideology, reappear in modified guise. Environmental determinism is given new life when the environment in question becomes the perceived environment in peoples' heads. Biological determinism is renewed by plausible genetic mechanisms to explain altruistic behaviour.

The renewable quality of older determinisms thus rests in part on new scientific discoveries but also in part on the need for correctives. The pendulum-like oscillation in intellectual thought found in many disciplines often excludes much that was insightful in the previous swing of ideas. To break through the existing walls of hostility or indifference, neo-determinists state their case with a vigour or passion that might be unnecessary in a less contentious environment. In a more reasonable climate of inquiry, determinist theory would seek to identify those areas of the human environment in which the anticipated cause and effect linkages are strongest and their explanations most parsimonious.

Partial theories persist not merely as correctives but because they illuminate, often in fundamental ways, some significant area of the human environment. Yet in so doing they still retain much of the simplistic attraction of one-dimensional explanation, the archetype of which is the classic statement from the Communist Manifesto: 'The history of all hitherto existing society is the history of . . .' Theories that conclude the sentence with a word or two appeal to a deep human need to simplify the complexity of the world.

It is likely that interactive theories will persist as well but it is too early to tell. With the exception of a few conceptual triads, interactive theories are young, less than twenty-five years old and the global models are but a decade old. Their persistence remains to be demonstrated. To illustrate this persistence of theory (Kates, 1983; Burton and Kates, 1986), I use the most frequently disproved and resurrected theory of the human environment: Malthus's theory of the resource–population relationship. (See Figure 2.1.) The remarkable thing about Malthus's *Essay on Population* is not that it fits the cynical definition of a classic given by Petersen (1979): 'a work that everyone cites but no one reads,' but that, despite its repeated refutation, both in theory and observation, it arises anew in each successive intellectual generation by successive expansion of the numerator of resources and the denominator of population.

In 1798, the numerator of the Malthusian equation was solely food and agricultural land, not because Malthus was ignorant of other natural resource needs, but because food requirements so dominated other needs. By the 1850s, however, the imperatives of food requirements were expanded to other energy and material resources, marked by the classic volume of Jevons (1865) in Britain on the coal question, at the same time as national censuses made possible expansion of the denominator as well. In the United States the counterpart development might have been the plea of the American Association for the Advancement of Science (AAAS) in 1873 and 1890 (Proc. AAAS, 1890) for forest conservation. Food as a concern, however, does not disappear; it persists but on a wider population scale.

The post-war United States would mark a new turning point with the Paley Commission report (President's Materials Policy Commission, 1952), discounting fears as to resource scarcity and

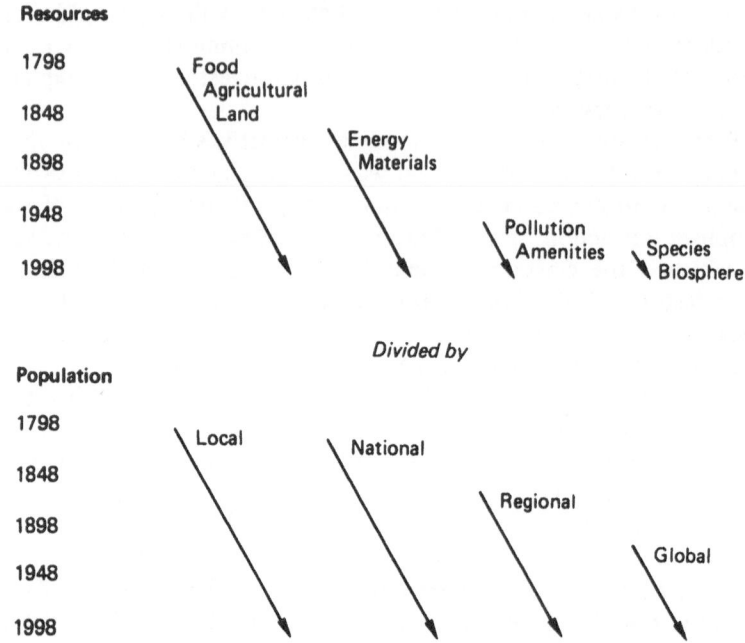

Figure 2.1   Expanding the Malthusian equation: resources divided by
            population

laying the groundwork for a new definition of the Malthusian
numerator involving amenity resources and the pollution-absorbing
capacity of the environment. The Stockholm Conference on the
Environment in 1972 would enlarge these concerns to global scale
centred on the biosphere and the basic life-support system of
biogeochemical cycles (Holdgate, Kassas and White, 1982; Clark &
Munn, 1986). The final extension is the current concern with extinction
(Regenstein, 1975; Myers, 1979; Ehrlich and Ehrlich, 1981) and
predictions of massive species destruction over the next several
decades.

But the earlier definitions by no means disappear; they persist. In
the 1970s the concern with food adequacy gained new life in the
context of the Sahelian drought. The concern with energy and
material resource adequacy was revived by the massive increase in
commodity prices during the decade and the recognition of the
limitations of oil reserves. As with the earlier Malthusian predictions,
the neo-Malthusian predictions also fail. A classic case, of course, is

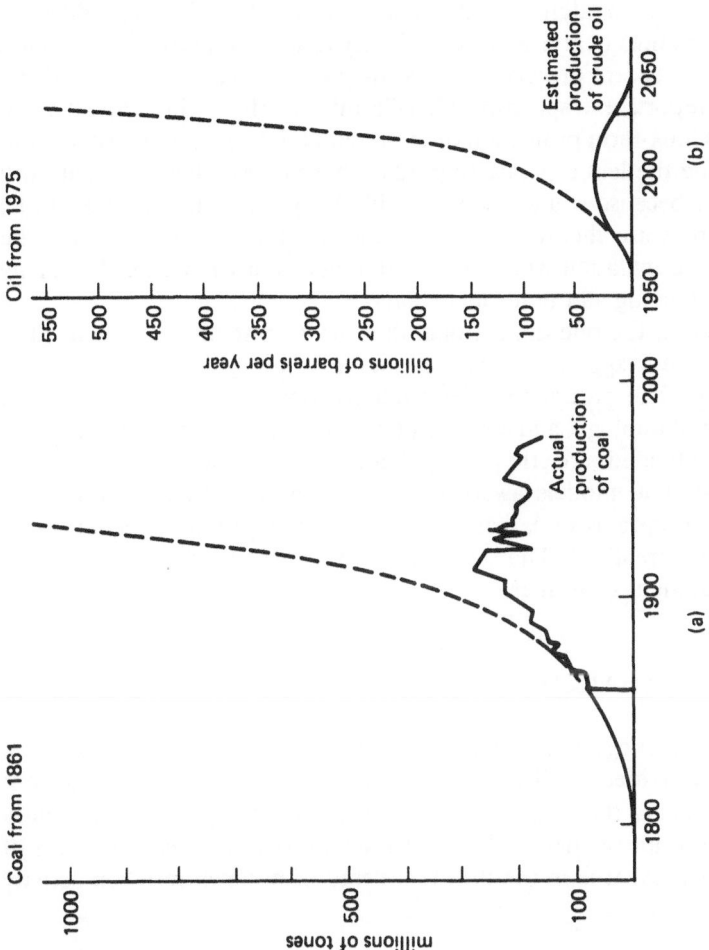

Figure 2.2   Extrapolating coal use from 1861, oil use from 1975

Sources: Jevons, 1865; United States, Council on Environmental Quality, 1980.

Jevons (1865) on the coal question (Kates, 1983b). An extrapolation of the growth in coal use from the time he wrote his book is shown in Figure 2.2(a) along with the actual course of British coal production. These figures bear a striking similarity to those represented in Figure 2.2(b), a recent projection concerning oil use taken from the *Global 2000* reports extrapolating 1950–75 oil growth continued with Hubbert's long-term projection of expected United States oil production.

None the less it is safe to predict that neo-Malthusian debate will persist because some resources will always be in limited supply, all resources are theoretically finite and, over the next 100 years, the earth's population will surely double and redistribute itself in many ways (barring all-out nuclear wear).

In sum, the one-dimensional theories persist, the reports of their death are exaggerated; they re-emerge in new guises with surprising vitality. The partial theories evidence power, provide explanation and methodology, and resist refutation in limited domains; only when over-extended are they truly contentious. The global concepts, systems and models await elaboration: new concepts, more data, better programs or larger computers. Their promise, seemingly, is always unrealised. The search for distinctive theories of the human environment continues.

II THE SEARCH FOR THEORY

My own suggestions for search follow five overlapping lines of inquiry divided by different levels of generalisation and abstraction: metatheory, selected theory, extended theory, prescriptive theory and descriptive theory. The first four are only suggestions, a list of possibilities. It is in the fifth that some of my own research activity falls.

Metatheory

A fierce territoriality governs the social behaviour of science. Never is it more visible than in the outcries and war-cries of anthropologists deploying to resist sociobiology or economists defending the dollar against the BTU (British Thermal Unit). There is a territoriality to theory as well, and one that might be used to advance rather than to retard human environment theory. This review has tried to demonstrate a truism about theory, namely that the ones that persist, the 'born again' theories, have two common characteristics. They are simplistic, yet embody truth. Within limited domains, they explain things, they have power, they speak to partial truths that most people know about. They appeal to our cognitive limitations in coping with complexity. They are satisfying.

What confuses is when we extend their domains – when we try to explain revolutionary movements as Oedipal conflicts or Kwakiutl potlatch give-aways as factors of production. We think of these extensions as over-simplifications, interesting thoughts, but surely not all of the story. Indeed, we know that Oedipus does not even explain all parental conflict, nor the factors of production the way capitalism works. But they do explain a great deal. They are closer to the reality that the theories arose from, and thus better fit the domain.

Physical science has accepted for some time differing constructs of similar phenomena, for example, wave and particle theories of light. These are used to supplement each other as in examining a painting from differing vantage points. Is there some way to order theories relevant to the domains in which they are most enlightening and to learn to shift vantage points to enhance understanding or, better yet, to answer questions? There is potential for metatheory, to taxonomise and thus divide the theoretical estate of the human environment and to match the one-dimensional and partial theories to their productive issues. The task of constructing metatheory could begin with both the theories and the problems they seek to explain or solve. Beginning with theories, we ask what are the phenomena that they best and most directly explain. For the problems, we ask what are the theories that appear most relevant. Then, with our two lists, we examine the overlap and ask if there is an underlying structure and whether we can state its principles. If we can, we have a metatheory.

Selected theory

For some theoreticians, for example, Marvin Harris, a particularly annoying adversary to engage is the elusive, but common, eclectic:

> I have found most of my colleagues think it unscientific to make an explicit commitment to one or another research strategy. They listen attentively to the claims and counterclaims of each strategic option, but they refuse to acknowledge a need for adopting one

strategy to the exclusion of the rest. Does not science oblige researchers to keep an open mind? In the human sciences this seems like an especially good idea. With so many extreme positions being advocated at once, isn't it likely that they all have something worthwhile to offer? It scarcely seems possible that only one out of the lot could have a decisive advantage over all the others. Isn't it obvious that prudent scientists will not follow any particular strategy but rather will reserve the right to pick and choose among the whole spectrum of epistemological and theoretical principles according to whatever seems to work best for the particular problem they happen to be working on?

Those who believe that prudence and common sense demand that one must avoid a commitment to any particular research strategy fail to realize that such a belief constitutes a commitment to a definite research strategy – the strategy of eclecticism. This strategy scarcely qualifies as prudent or scientifically sensible. By picking and choosing epistemological and theoretical principles to suit the convenience of each puzzle, eclecticism guarantees that its solutions will remain unrelated to each other by any coherent set of principles. Hence eclecticism cannot lead to the production of a corpus of theories satisfying the criteria of parsimony and coherence. Rather, eclecticism is a prescription for perpetual scientific disaster: middle-range theories, contradictory theories, and unparsimonious theories without end. (Harris, 1979; pp. 287–8).

What is intriguing about his vehemence is that despite his protestations, Harris is himself an example of what I would label as a selective eclectic. In his case, he seeks to revive Marx by accommodating Malthus and to accept materialism and reject dialectics. Selective eclecticism is an interesting strategy.

One might begin with selections from the persuasive partial theories, particularly those of the late eighteenth and nineteenth century giants Darwin, Freud, Marx and Malthus. In Figure 2.2, I have tried to sketch some relationships between them and some current efforts at synthesis. With the unbroken line I link the acknowledged intellectual debt of Darwin to Malthus and of Marx to Darwin. I know less of Freud's writings. With broken lines I have identified some of the recent attempts at synthesis, or of selective eclecticism. Marcuse (1966) and his student Leiss (1972), and Heilbroner (1974) (as well as others) have tried to integrate Freud and Marx; Harris (1979) and Daly (1971) have tried to bring Malthus

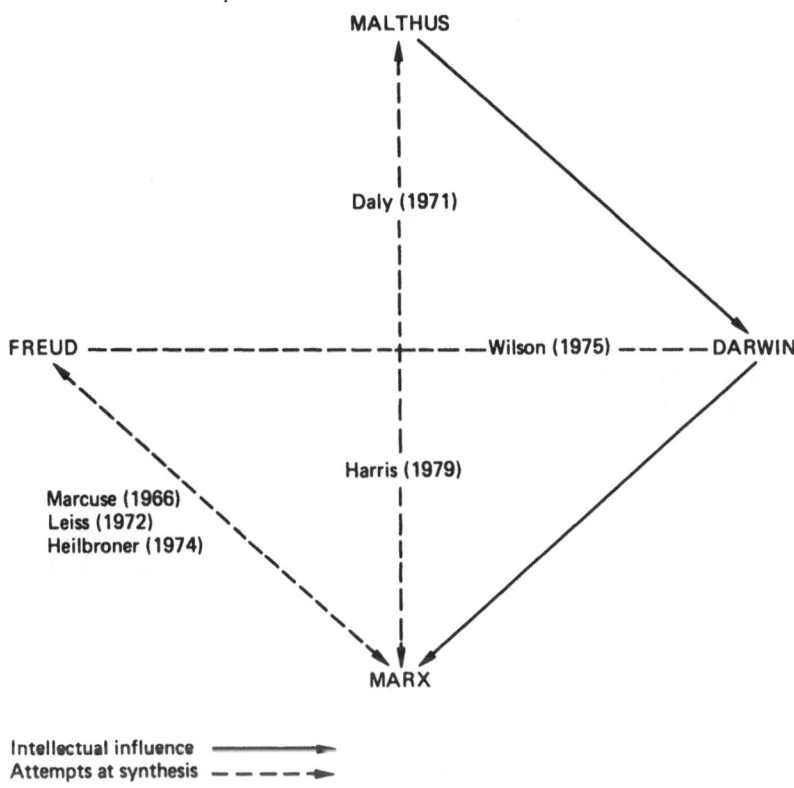

Figure 2.3   A selective eclectic of the nineteenth century

into Marxist theory; and Wilson (1975) and the sociobiologists are in one sense trying to relate Darwin and Freud by way of modern genetics. Except for Heilbroner (1974) and Daly (1971) none of these selective efforts have used the major problems of the human environment as a starting point and some effort at selective eclecticism seems warranted.

Extended theory

In sociology one can contrast Merton's (1968) theories of the middle-range with Mills's (1959) descriptions of grand theory. Is there something we can call theories of the extended range, theories that are extensions of well-examined middle-range theory? One way to explore such potential is to try to extend natural and technological hazard theory to resources. A good integrated hazard/resource theory would come close to a distinctive theory of the human environment.

It is hazard theory rather than resource theory that could serve as a base for such extension. For the last twenty-five years, natural resource theory has been a transfer theory. Resources were to be allocated as the land component in the factors of production of private enterprise, and a simulated market analysis for profitability was to be developed for public allocation in the form of cost–benefit analysis. At the same time, natural hazard theory (Burton, Kates and White, 1978; Hewitt, 1983) and (later) technological hazards (Kates, Hohenemser and Kasperson, 1985) attracted a wide interdisciplinary following in which natural and social scientists combined to ask such questions as: is disaster an inevitable concomitant to the human use of the earth, are hazard losses increasing in both industrialised and developing countries, or how is it that people persist and even prosper in areas of high recurrent natural hazards? Variants of these questions are also being asked in relationship to technology and its hazards with a strong integration of bio-physical and social behavioural science. Central to both lines of research is the recognition of hazard as a joint product of nature, society and technology and as inherent to the use of resources; risk always accompanies benefit. The resulting theoretical positions are still evolving and are in dispute but it is a lively and productive area and capable of extension.

Prescriptive theory

A final suggestion for theory-building is to create a normative theory – to cloak in the lawlike mantle of theory some desired state of the human environment. The argument for doing so might begin with two observations: first, that the choices between existing theories may be irreconcilable. No scientific research programme is likely to create by itself an overwhelming case for one set of choices over another. Second, that all of the major problems of the human environment have multiple solutions (including doing nothing), each of which, conceivably, might lead to similar outcomes. Thus there may be no objective method for choosing between theories or for using such theories to indicate desirable social actions.

Therefore we might argue that, instead of seeking a 'true' or 'powerful' theory, we should construct a theory directed towards some desired ends. A plausible, compelling, but consensually-accepted theory of the human environment might become a self-fulfilling theory. If people and nations accepted it, they would set about acting upon it and thereby vindicate it.

One can only speculate about what such a theory would look like but a few qualities can be suggested. First, it should be scientifically plausible and, indeed, given the secular religious rule of science, well draped in a scientific mantle. Second, it should have an aspect of both historic inevitability and Utopian goals. Both Christianity and Marxism seem to have commanded support by the recognition of freedom as necessity combined with Utopian futures. Third, unlike Christianity or Marxism, it should be incrementally implementable, not requiring instantaneous or immediate revision in social behaviour. Fourth, and perhaps most important, it must speak significantly to the great questions of the human environment: Malthusian scarcity, the global division of rich and poor, the social control of technology (Kates, 1983b).

My own candidate for such a theory would be a theory of the great climacteric, a prescription for the way we should behave in the final third of the transition from the industrial revolution's surge of population to the steady-state of the mid-twenty-first century. A climacteric is, in the words of the *Oxford English Dictionary*, 'a critical period in human life' and 'a period specially liable to change in health and fortune'. Ian Burton and I use it not for individuals, but to characterise the global opportunity and risk in achieving a just and sustainable human environment (Burton and Kates, 1986). Given the prospect of fairly providing for at least a doubled population in a world with a current arms budget of $800 billion per annum, the lifetime future of a child born today is indeed subject to 'change in health and fortune'.

We take symbolically, as the period of the climacteric, the date of Malthus's first edition of the *Essay on Population* and 2048 (the mid-century optimistic forecast for a steady-state population), a period of 250 years. In our view a just and sustainable global human environment of 2048 will have at least five principal characteristics.

First a major, successful effort to reduce the superpowers' and other powers' nuclear arms and to limit their proliferation will have been made, with at least partial success in limiting the share of gross world product spent on armaments. Whether such a shift will come only as a result of a nuclear exchange which stops short of full-scale war is an open question. We are not optimistic that all use of nuclear weapons can be prevented before the world comes to its senses.

Second, by 2048 the global population will have clearly begun to level off. Over large areas of the world the rate of increase will have slowed down to that prevailing in the most stable countries (in population terms) of today.

Third, the wealth of nations will be much more uniform than today. A pattern for nations analogous to the distribution of wealth among people in today's more egalitarian states, like Sweden or Britain, seems possible and, indeed, essential. This would mean a very few wealthy states (perhaps benefiting from fortuitous endowment with a national resource which is in high demand), a small number of very poor states (the last to achieve population stability) and a very large 'middle class' of nations distributed in a narrow band around the mean.

Fourth, there will be a much greater degree of social discipline and control, especially in those states where it is now lacking. These are of two types – developing countries where the apparatus of state bureaucracy is not yet fully established, and the wealthier industrial nations of today that are attempting to combine a high degree of social discipline in some areas while preserving a high degree of freedom of choice in others. To much conventional Western thinking, the future, more disciplined and bureaucratically controlled society sounds like a society without freedom. The choices of individuals and organisations would be everywhere constrained. This is rather a challenge for the new society of 2048, that is, to find ways of management that ensure the necessary control while preserving the individual sense of freedom, and avoiding a fatal overweight of bureaucracy.

Lastly, there would be a new relationship with the environment. Decisions would be guided by a concern for the long-term viability of the biosphere, and a recognised set of 'rights' for the natural environment, especially other living oganisms, would be included. A believable theory of the necessity and inevitability of these, or similar requirements, for a just and sustainable world would increase our chances of creating such a world.

Descriptive theory

Descriptive theory attempts to explain 'what is' as opposed to prescriptive theory which attempts to explain 'what ought to be'. While the distinction frequently becomes muddled in practice, descriptive theory often attempts to explain some observed examples of human environments or resource and environmental behaviour. Sometimes the case examples serve inductively to derive theory, generalised from their particulars; sometimes to test theory through predictions about the particular; but more often than not, to illustrate theory, basically pithy anecdotes, to make a point.

In recent years, I have worked with colleagues on three sets of case studies. These studies seek to create theory inductively, trying to generalise from common characteristics of the particular. The case studies vary widely but have one common characteristic: nature, as well as society and technology, appears a priori to be an important element. In each case a question is posed, calling for explanation.

(1) What are the causes of least-developed nations?

Initially twenty-five and then, by various other definitions, upwards of fifty nations have been ranked as least-developed or the 'poorest of the poor' by international organisations in order to facilitate special efforts at trade or aid. Len Berry and I, with other colleagues (Berry and Kates, eds, 1980), were drawn to this problem on examining Figure 2.4, a map of the original set of least-developed nations. We noted that their distribution (primarily in the Sahel–Savanna regions of Africa and the mountain fringe of Asia) appeared non-random and were conceivably linked to geographical location.

To explain the causes of qualification as 'least-developed', we explored a wide range of theoretical positions, but essentially of three types: theories that emphasise the inherent characteristics of the nations involved (unfortunately endowed, young, inept), theories that emphasise processes of underdevelopment, and theories that challenge the very concept of least development, seeing it as a cultural artefact of the inappropriate measurement scales employed. In addition we explored a special theory of leastness, that is, randomness: 'after all, someone must be last'. In the end, Berry and I concluded that least development arises from interaction of two sets of relationships – economic and environmental marginality. By location and wealth, least-developed nations are on the periphery of the periphery (in the sense of core-periphery theory). This periphery is characterised by physical isolation, arid or semi-arid climate or high mountain relief and climate. Environment alone does not create poverty, but it can make the poor, poorest.

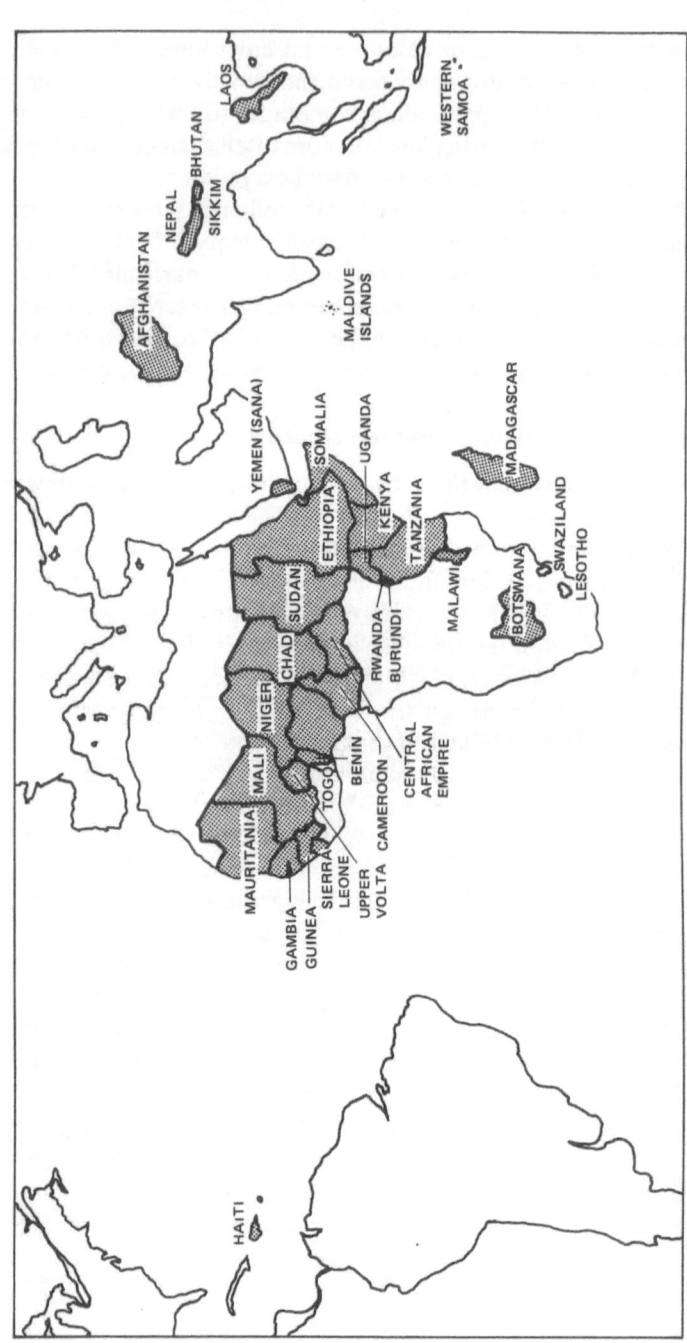

Figure 2.4 **Least-developed nations**

(2)   What is the link between reduced variability and catastrophe?

For two decades, two independent research groups, one linked to hazard research and the other to ecological management, have collected case studies in which some undesired variation of nature (storms, floods, weeds, pests, disease) or of technology (accidents, hazards) are reduced or eliminated, yet losses from such phenomena are not reduced or are even increased. A common characteristic of such situations is that of lessening and catastrophe, that is, there are fewer accidents, failures, events, outbreaks, but when they do occur they are more serious, sometimes devastatingly so. See Table 2.3.

Table 2.3   Case studies of fewer failures and larger losses

NATURAL HAZARDS	
Biological:	Budworm
	Fire
	Gonorrhea
	Malaria
Geophysical:	Droughts
	Floods
	Tropical cyclones
NATURAL RESOURCES	
Renewable:	Forest
	Salmon
	Savanna
TECHNOLOGICAL HAZARDS	
	Fire
	Auto
	TV

The two classic examples of this apparent relationship are flood control and fire control. For example, building levees to keep water out of human settlements is effective for a certain level of riverflow; above that level levees are overtopped, resultant damage being much greater than would have occurred in the absence of the levee. Similarly, modern fire-fighting in protected nature reserves suppresses small fires, allowing combustible material to accumulate and leading to less frequent but far more severe fires. There are, in addition, over a dozen well-documented case studies of this relationship. Currently, with C. S. Holling, I am exploring the common characteristics of such case studies. It is not clear yet whether they should be viewed as a common occurrence of human intervention in natural systems or as exceptional cases. Indeed, over time, lessening of impact may occur (Bowden *et al.*, 1981). This in turn relates to larger issues of 'doomsday theory', the repeated warnings of environmental doomsdays and the rarity with which they actually occur.

(3)   What are the causes of millennial long waves in human occupancy?

In some places, over long periods of time, doomsday does occur. This is only perceived when place is held constant and time is extended beyond conventional periods of historical analysis. Using numbers of people as a measure of human occupancy, waves of growth and decline rather than simple cumulative growth appear common at scales of millennia in fixed regions. Scattered evidence for the existence of these long waves is found in prehistorical and historical reconstruction of human occupancy in such diverse regions as the central plateau of Mexico, the northern coast of Peru, the southwestern United States, or the valleys of the Nile, Yangtze, and Tigris–Euphrates rivers.

Our work to date has centred on what appeared to us as the best reconstruction to date, that of the Tigris–Euphrates floodplain, based on the exceptional work of R. McAdams (1965, 1972, 1981). He has undertaken comprehensive archaeological surveys of the region and related these to the many archives of clay tablets and to historical writing (for example, Walters, 1970). Expressed as a summary population curve, two and a half waves of growth and decline have been experienced in the region over the past 6000 years. (See Figure 2.5.) Our current research programme (with P. Burke, T. Gottschang, D. L. Johnson and B. L. Turner III) employs a computer model of the best archaeological–historical description of life and livelihood on the Tigris–Euphrates floodplain (Johnson and Gould, 1984) which reproduces millennial long waves. We have expanded this work, by updating the Tigris–Euphrates data to the present, and have added reconstructions of the Nile Valley, Egypt; the Basin of Mexico; and the Central Mayan Lowlands, Belize, Guatemala and Mexico.

Each of these sets of case studies initially emphasises nature, society and technology. Place and natural setting are kept constant in the Tigris–Euphrates to allow for the truly rare events of nature to occur in the six millennia of the study. Technology is the

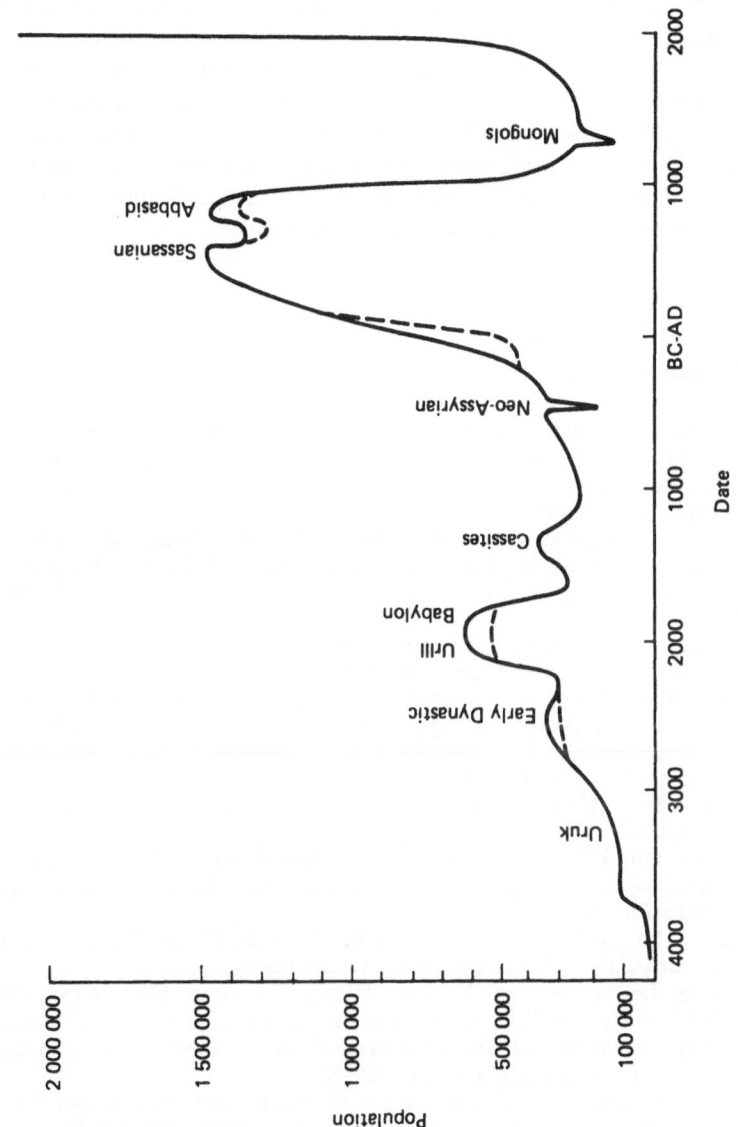

Figure 2.5 Historical population levels on the Tigris–Euphrates floodplain (a speculative reconstruction). (Dashed lines show alternative reconstructions for certain periods that are certainly plausible but that have been tentatively rejected)

major intervention in all the case studies of resource and hazard management. And nationhood (as a surrogate for society) is the central unit in the least developed. The common theme, however, is the observation of the interactive behaviour of all three factors and the generalisation from these. However it is still not clear whether new theory can arise from such disparate empirical questions any more than it can arise from reviews of such different theoretical perspectives. The search for theory continues.

References

American Association for the Advancement of Science (1890) *Proceedings of the American Association for the Advancement of Science* 39: 28.

Ayres, Robert U. (1978) *Resources, Environment, and Economics* (New York: John Wiley).

Berry, Leonard and Robert W. Kates (eds) (1980) *Making the Most of the Least: Alternative Ways to Development* (New York: Holmes & Meier).

Bookchin, Murray (1971) *Post-Scarcity Anarchism* (Palo Alto, Calif: Ramparts Press).

Boulding, K. E. (1981) *Ecodynamics: A New Theory of Societal Evaluation* (Beverly Hills, Calif: Sage Publications).

Bowden, Martin J. *et al.* (1981) 'The Effect of Climate Fluctuations on Human Populations: Two Hypotheses', in *Climate and History: Studies in Past Climates and Their Impacts on Man*, edited by T. M. L. Wigley *et al.* (Cambridge: Cambridge University Press).

Buckle, H. T. (1899) *A History of Civilization in England*, 3 vols (London: Longman Green).

Burton, Ian, Robert W. Kates and Anne Kirkby [Whyte] (1974) 'Geography', *Interdisciplinary Environmental Approach* (Costa Mesa, Calif: Educational Media Press).

Burton, Ian, Robert W. Kates and Gilbert F. White (1978) *The Environment as Hazard* (New York: Oxford University Press).

Burton, Ian and Robert W. Kates (1986) 'The Great Climacteric, 1798–2048', in Robert W. Kates and Ian Burton (eds), *Geography, Resources, Environment, Vol. II: Themes from the Work of Gilbert F. White* (Chicago: University of Chicago Press) pp. 339–60.

Butzer, Karl W. (1982) *Archaeology as Human Ecology: Method and Theory for a Contextual Approach* (Cambridge: Cambridge University Press).

Caldwell, Lynton K. (1969) 'Health and Homeostasis as Social Concepts: An Exploratory Essay', *Diversity and Stability in Ecological Systems*, Brookhaven Symposia in Biology No. 22.

Cannon, Walter B. (1932) *The Wisdom of the Body* (New York: W. W. Norton).

Caplan, Arthur L. (1978) *The Sociobiology Debate: Readings on the Ethical*

and Scientific Issues Concerning Sociobiology (New York: Harper & Row).
Clark, W. and T. Munn (eds) (1986) *Sustainable Development of the Biosphere* (Cambridge: Cambridge University Press).
Cole, H. S. D. *et al.* (1973) *Models of Doom* (New York: Universe Books).
Commoner, Barry (1971) *The Closing Circle* (New York: Alfred A. Knopf).
Covello, Vincent and Mark Abernathy (1983) 'Risk Analysis and Technological Hazards: A Policy-Related Bibliography', in C. Whipple *et al.* (eds), *Technological Risk Assessment* (Winchester, Ma: Sitjthoff & Noordhoff).
Cronbach, Lee J. (1975) 'Five Decades of Public Controversy Over Mental Testing', *American Psychologist* 30, pp. 1–14.
Daly, Herman E. (1971) 'A Marxian–Malthusian View of Poverty and Development', *Population Studies* 25, pp. 25–37.
Diener, Paul and Eugene R. Robkin (1978) 'Ecology, Evolution and the Search for Cultural Origins: The Question of Islamic Pig Prohibition', *Current Anthropology* 19, pp. 493–540.
Douglas, Mary (1966) *Purity and Danger: An Analysis of Concepts of Pollution and Taboo* (London: Routledge & Kegan Paul).
Dubos, René (1965) *Man Adapting* (New Haven: Yale University Press).
Duncan, Otis Dudley (1964) 'Social Organization and the Ecosystem', in R. E. L. Furis (ed.), *Handbook of Modern Sociology* (Chicago: Rand McNally).
Ehrlich, Paul and Anne Ehrlich (1981) *Extinction: The Causes and Consequences of Disappearing Species* (New York: Random House).
Eliade, Mircea (1959) *The Sacred and the Profane: The Nature of Religion* (New York: Harcourt, Brace & World).
Ellen, Roy (1982) *Environment, Subsistence and System: The Ecology of Small-Scale Social Formations* (Cambridge: Cambridge University Press).
Ellul, Jacques (1973) *The Technological Society* (New York: Alfred A. Knopf).
Firey, Walter (1960) *Man, Mind and Land* (Glencol, Ill: The Free Press).
Fisher, Anthony C. (1981) *Resource and Environmental Economics* (Cambridge: Cambridge University Press).
Forde, C. Darryl (1963) *Habitat, Economy and Society*, 5th edn (New York: E. P. Dutton).
Freeman, Christopher, John Clark and Luc Soete (1982) *Unemployment and Technical Innovation: A Study of Long Waves and Economic Development* (London: Francis Pinter).
Giedion, S. (1948) *Mechanization Takes Command* (New York: Oxford University Press).
Glacken, Clarence J. (1967) *Traces on the Rhodian Shore* (Berkeley: University of California Press).
Goodman, Daniel (1975) 'The Theory of Diversity–Stability Relationships in Ecology', *The Quarterly Review of Biology* 50, 237–66.
Griffin, Susan (1978) *Woman and Nature: The Roaring Inside Her* (New York: Harper and Row).
Hamilton, W. D. (1963) 'The Evolution of Altruistic Behavior', *American Naturalist* 97, 354–6.
Hardin, Garrett (1968) 'The Tragedy of the Commons', *Science* 157, pp. 1243–8.

Harris, Marvin (1966) 'The Cultural Ecology of India's Sacred Cattle', *Current Anthropology* 7, 51–9.

Harris, Marvin (1979) *Cultural Materialism: The Struggle for a Science of Culture* (New York: Random House).

Heilbroner, Robert (1974) *An Inquiry into the Human Prospect* (New York: W. W. Norton).

Hewitt, K. (ed.) (1983) *Interpretation of Calamity from the Viewpoint of Human Ecology* (Boston: Allen & Unwin).

Holdgate, Martin, Mohammed Kassas, and Gilbert F. White (1982) *The World Environment 1972–1982: A Report to the United Nations Environment Programme* (Dublin: Tycooly International).

Holling, C. S. (1973) 'Resilience and Stability of Ecological Systems', *Annual Review of Ecology and Systematics* 4, pp. 1–23.

Holling, C. S. (ed.) (1978) *Adaptive Environmental Assessment and Management* (Chichester, United Kingdom: John Wiley).

Huntington, Ellsworth (1965) *Civilization and Climate* (New Haven: Yale University Press).

James, Patricia (1979) *Population Malthus: His Life and Times* (London: Routledge & Kegan Paul).

Jevons, W. S. (1865) *The Coal Question: An Inquiry Concerning the Progress of the Nation on the Probable Exhaustion of Coal* (London: Macmillan).

Johnson, D. L. and H. Gould (1984) 'The Effect of Climate Fluctuations on Human Populations: A Case Study of Mesopotamian Society', in *Climate, Environment and Society*, edited by A. K. Biswas (Oxford: Pergamon Press).

Kates, Robert W. (1983a) 'Part and Apart: Issues in Humankind's Relationship to the Natural World', in F. Kenneth Hare (ed.), *The Experiment of Life: Science and Religion* (Toronto: University of Toronto Press) pp. 151–80.

Kates, Robert W. (1983b) 'The Human Environment: Penultimate Problems of Survival', Natural Hazards Research and Applications Center, Special Publications No. 6 (Boulder: University of Colorado Institute of Behavioral Science).

Kates, Robert W. and Jeanne X. Kasperson (1983) 'Comparative Risk Analysis of Technological Hazards (A Review)', *Proceedings of the National Academy of Sciences, USA* 80, pp. 7027–38.

Kates, Robert W., Christoph Hohenemser and Jeanne X. Kasperson (eds) (1985) *Perilous Progress: Technology as Hazard* (Boulder, Co: Westview Press).

Kropotkin, Peter (1902) *Mutual Aid: A Factor of Evolution* (New York: Doubleday).

LePlay, Frédéric (1879) *La Méthode Sociale: Abrigé des Ouvriers Européens*, 2nd edn (Tours: A. Mame).

Leiss, William (1972) *The Domination of Nature* (Boston: Beacon Press).

Leontief, Wassily (1977) *The Future of the World Economy* (New York: Oxford University Press).

Lewontin, R. C., Stephen Rose, and Leon J. Kamin (1984) *Not in Our Genes: Biology, Ideology and Human Nature* (New York: Pantheon Books).

Lovelock, J. E. (1979) *Gaia: A New Look at Life on Earth* (Oxford: Oxford University Press).
Lovins, Amory B. (1976) 'Energy Strategy: The Road Not Taken?' *Foreign Affairs* Oct. 1976, pp. 65–95.
Lowenthal, David (1958) *George Perkins Marsh: Versatile Vermonter* (New York: Columbia University Press).
McAdams, Robert (1965) *Land Behind Baghdad: A History of Settlement on the Diyala Plains* (Chicago and London: University of Chicago Press).
McAdams, Robert (1981) *Heartland of Cities: Surveys of Ancient Settlement and Land Use on the Central Floodplain of the Euphrates* (Chicago and London: University of Chicago Press).
McAdams, Robert and Hans J. Nissen (1972) *The Uruk Countryside: The Natural Setting of Urban Societies* (Chicago and London: University of Chicago Press).
Marcuse, Herbert (1966) *One-Dimensional Man* (Boston: Beacon Press).
Marsh, George Perkins (1965) *Man and Nature* (Cambridge, Mass: The Belknap Press of the Harvard University Press).
Marx, Leo (1964) *The Machine in the Garden: Technology and the Pastoral Ideal in America* (New York: Oxford University Press).
Meadows, Donella, *et al.* (1972) *The Limits to Growth: A Global Challenge* (New York: Universe Books).
Meadows, Donella, John Richardson and Gerhart Bruckmann (1982) *Groping in the Dark: The First Decade of Global Modelling* (Chichester, United Kingdom: John Wiley).
Meggars, B. J. (1954) 'Environmental Limitations on the Development of Culture', *American Anthropologist* 56, pp. 801–23.
Merton, Robert K. (1968) *Social Theory and Social Structure* (New York: The Free Press).
Miller, James Grier (1978) *Living Systems* (New York: McGraw-Hill).
Mills, C. Wright (1959) *The Sociological Imagination* (New York: Oxford University Press).
Myers, Norman (1979) *The Sinking Ark: A New Look at the Problem of Disappearing Species* (Oxford: Pergamon Press).
Odum, Howard T. (1976) *Energy Basis for Man and Nature* (New York: McGraw-Hill).
Odum, Howard T. (1983) *Systems Ecology: An Introduction* (New York: John Wiley).
Passmore, John (1974) *Man's Responsibility for Nature* (New York: Charles Scribner's Sons).
Petersen, William (1979) *Malthus* (Cambridge, Mass: Harvard University Press).
President's Materials Policy Commission (1952) *Resources for Freedom* (Washington: United States Government Printing Office).
Ray, John (1759) *The Wisdom of God Manifested in the Works of Creation* (London: Rovington, Ward, Richardson).
Regenstein, Louis (1975) *The Politics of Extinction: The Shocking Story of the World's Endangered Wildlife* (New York: Macmillan).
Sachs, Ignacy (1977) 'Civilization Project and Ecological Prudence', *Alternatives III*, pp. 1–18.

Schumacher, E. F. (1973) *Small is Beautiful* (New York: Perennial Library).

Semple, E. (1911) *Influences of the Geographic Environment: On the Basis of Ratzel's System of Antrhopo-geography* (New York: Henry Holt).

Shaw, William H. (1979) 'The Handmill Gives You the Feudal Lord: Marx's Technological Determinism', *History and Theory* 18, pp. 155–76.

Simon, Julian L. (1981) *The Ultimate Resource* (Princeton: Princeton University Press).

Simoons, F. J. (1979) 'Questions in the Sacred-Cow Controversy', *Current Anthropology* 20, pp. 467–93.

Spencer, Herbert (1910) *The Principles of Sociology*, vols 1–3 (London: D. Appleton).

Sproul, Barbara C. (1979) *Primal Myths: Creating the World* (San Francisco: Harper & Row).

Starr, Chauncey (1969) 'Social Benefit versus Technological Risk', *Science* 165, pp. 1332–38.

Thomas, Jr William L. (ed.) (1956) *Man's Role in Changing the Face of the Earth* (Chicago: The University of Chicago Press).

Walters, Stanley D. (1970) *Water for Larsa: An Old Babylonian Archive Dealing with Irrigation*, Yale Near Eastern Researcher, 4 (New Haven and London: Yale University Press).

Ward, Barbara and René Dubos (1972) *Only One Earth: The Care and Maintenance of a Small Planet* (London: André Deutsch).

White, Jr Lynn (1967) 'The Historical Roots of Our Ecological Crisis', *Science* 155 10, March 10, pp. 1203–7.

White, Jr Lynn (1973) 'Continuing the Conversation', in I. G. Barbour (ed.), *Western Man and Environmental Ethics* (Reading, Mass: Addison-Wesley).

Williams, Raymond (1976) *Keywords: A Vocabulary of Culture and Society* (New York: Oxford University Press).

Wilson, E. O. (1975) *Sociobiology: The New Synthesis* (Cambridge, Mass: Harvard University Press).

Winner, Langdon (1978) *Autonomous Technology: Technics-out-of-control as a Theme in Political Thought* (Cambridge, Mass: MIT Press).

Worster, Donald (1977) *Nature's Economy: The Roots of Ecology* (San Francisco: Sierra Club).

3 Reinterpreting the Concept of Development from a Science and Technology Perspective

Francisco R. Sagasti

I INTRODUCTION

The major advances in science and technology at the world level over the last thirty years are making it necessary to reinterpret the concept of 'development' and to offer explanatory schemes that incorporate explicitly the process of knowledge generation. However it is surprising how little attention development economists have given to this theme, even though there are a few notable exceptions to this general rule, with some development economists having turned their attention to technological issues in a partial way in recent times.[1]

This essay constitutes a preliminary attempt to incorporate explicitly into the conceptualisation of the development process some issues referring to science and technology. The starting point is a critique of the general model offered by George Basalla to explain the diffusion of Western science. This is followed by a brief description of the components of an alternative conceptual framework, and the outline for an explanatory scheme that would link the various components of the alternative model in an organic fashion. The present essay develops a line of work advanced by the author in other reports, and is part of a larger research project currently under way.[2]

II BASALLA AND THE DIFFUSION OF WESTERN SCIENCE

In a well-known paper George Basalla (1967) proposed a conceptual framework to explain the spread of Western science throughout the world. His model consists of three, partly overlapping, stages: in the first stage, the non-scientific or pre-scientific society of the developing world constitutes a source of problems for European science; in the second, there is an incipient development of what Basalla calls 'colonial science'; and the third stage is characterised by a struggle to establish an independent scientific tradition.

During the first stage, a few European scientists visit the new lands, explore and collect fauna and flora, study the geographical and physical characteristics of unexplored areas, and then return to their place of origin to complete their scientific work.

A dependent 'colonial science' emerges in the second stage. Natural history continues to be the main focus of interest and attention, but the range of scientific activities and problems studied begins to expand until it almost coincides with that of the colonial power. The colonial scientist is dependent in the sense that the sources of his education and training, the origin of the scientific traditions that he adheres to, the orientation of his activities, and the ways of obtaining recognition for his work are all defined in the metropolitan scientific power, and not in the country or region in which he lives and works.

The transition from the second to the third stage is more complex and difficult to characterise. Basalla suggests that the stage of colonial science contains, in embryonic form, some of the essential aspects of the third phase. During this transition, the colonial scientist – even though he still gets support from outside – begins to create institutions and traditions that eventually will be the base for an independent scientific culture. Thus, in the third stage, the colonial scientist is replaced by a scientist whose main allegiances lie within the frontier of the country in which he works.

Basalla's model has two important limitations: the use of the concept of diffusion or 'dissemination' of 'Western science' as the principal axis, without giving sufficient attention to the processes of 'absorption' and 'internationalisation' of the scientific activities at the local level; and the fact that it centres attention only on the diffusion of Western science, without examining the expansion of the technological base and the internationalisation of productive activities.

To privilege the concept of dissemination or diffusion entails adopting an Europocentric perspective, in which 'Western science', nurtured by different currents of speculative, theoretical and empirical thought that converge upon it, irradiates the whole world until it displaces the local 'pre-scientific' forms of thought. In reality, what happened in different regions of the world, each of them with their own traditions and culture, was a process of interaction between the imported scientific knowledge and the traditional modes of speculative thought. The permanence of non-scientific forms of speculative thought is a constant in the history of Africa, India, China, Latin America, the Middle East, and even Japan, and the interaction between the occidental view of the world and the traditional perspectives has taken a variety of forms.

For these reasons, rather than speaking of 'diffusion' it would seem to be necessary to refer to the 'diffusion, absorption, and reinterpretation' of modern science, while admitting that this is a process which is still under way, and that in the developing world it is still at an incipient stage. However it is necessary to recognise that this interaction process has been (and is) taking place rather slowly between members of a European elite and their local counterparts; that in regions where local culture did not advance sufficiently it is very weak; and that in many places there has been little interaction, but rather a superimposition of two different and independent forms of speculative thought: the scientific Western view and the traditional autochthonous perspective.

On the other hand, when examining the diffusion of modern science without taking into account the parallel process of dissemination, absorption and adaptation of Western techniques (in which there were complex and rich interactions between the Western and the autochthonous traditions), and without considering the internationalisation of European productive activities (which accompanied the expansion of the capitalist system at the world level), there is the risk of presenting a partial vision, in which the 'diffusion of Western science' is perceived as an independent phenomenon, conditioned only by its own internal logic.

III TOWARDS AN ALTERNATE CONCEPTUAL FRAMEWORK

In order to offer an alternative view of the emergence and diffusion of modern science in the developing countries, it is necessary to consider the process of generation, transmission and utilisation of knowledge in an integral way. To this end, it is possible to distinguish a set of elements and components which, together with their interrelations and a certain directionality, form an alternative conceptual framework.

The first component is the *evolution of speculative thought* which seeks to generate knowledge in order to understand natural and social phenomena, to propose explanations that give sense to human existence, and to generate kowledge about the nature of the physical and social world. The second component is the *transformation of the technological base* that provides every human group with a set of organised responses (techniques) to confront the challenges posed by the physical and social environment, and also with the criteria to select from among these responses. The third component is the *modification and expansion of productive activities*, which provide goods and services to satisfy the needs of a community and of the individuals who compose it. These three elements or components, considered in a dynamic fashion as currents in constant transformation, insert themselves into the social, cultural and political context of every human group. What characterises a society at a given time and place is the degree of development of these three currents, the way in which they relate to each other, the form in which they are linked with their homologues in other societies, and the specific nature of the interaction among these three currents and their environment.

Even while it is necessary to reject the Western view as the unique frame of reference for comparing the achievements of different societies, it is impossible to deny that, considering the success in the material and the intellectual spheres and its diffusion on the global scale, the Western vision of 'progress' (which took several centuries to form) dominates the present world and has become an implicit standard. Without going back to the origins of this perspective in the Hellenic world, in the period between the sixteenth and eighteenth centuries, there was a qualitative, radical and unprecedented change that led to a 'Westernisation' of the concept of the natural world and of man's perception of himself. This conception was characterised by an emphasis on instrumental rationality, which subjected all human activities to the criterion of 'efficiency', subordinated human creativity to the process of accumulation, and took away the 'sacred' character of the natural world, thus creating the conditions for Western man to act with impunity over the physical environment. In this way the concern for means or instruments gradually replaced the preoccupation with identifying aims and giving an ultimate sense to the problem of human existence.

We are at present immersed in a world of values, conceptual schemes, artefacts and social entities, constructed by Western man and his instrumental rationality, to the point that, as Garaudy (1981) says, the West has confiscated the universal. For this reason, and in order to rescue the universal in all its diversity from the predominance of the West, it is necessary to examine and understand the impact that the West has had on the rest of the world. This requires, in the first place, a study of the evolution of the knowledge generation process, of the transformation of the technological base, and of the expansion at the world level of the productive system which characterised Europe and North America. However in this analysis it is necessary to be *en garde* in order to avoid adopting an implicit Europocentric perspective.

The development of different civilisations and societies over the last few centuries should be seen as a complex whole, whose components are in continuous action and transformation, and in which a perspective – the Western one – came to influence all others. But, at the same time, these other cultures preserved their individuality, affected Western culture, and gave rise to new hybrid means of conceiving the world and relating to it (Braudel, 1975). The image of all civilisations and cultures of the world converging to the culmination and greater glory of the Western civilisation, implicit in the metaphor of different cultures as tributary rivers which converge on the sea of Western culture, must be rejected. In this sense, it is convenient here to highlight what Alvarez (1979) has stated:

Human history may be better described not as a movement of different peoples towards some convergent mythical future (although at different speeds and in distinct groups), but as experience of many discontinuous cultures, each in itself equally important as exhibiting the variability of products of human inventiveness, each crystallizing a system of meanings irreducible to the others. (p. 2).

Ortega y Gasset (1968) has argued along the same lines, with particular reference to techniques, when he opposes:

[the tendency] as spontaneous as excessive, reigning in our time, to believe that in the last analysis there is truly no more than one technique, which is the actual European–American technique, and that everything else was just clumsy babble towards it.

[It is necessary] to counteract this tendency, and to submerge the technique of the present time as one of the many in a vast and multiform panorama of human techniques, revaluing in this way the sense and showing how to each project and model of humanity there corresponds a particular technique. (p. 77).

When discarding the perspective of Western civilisation as the frame of reference, in order to appreciate the march of other cultures, there still remains the problem of imparting a direction for the process of social evolution, which would act as a backdrop for any comparison. For this purpose, it appears adequate to accept the arguments of Wertheim (1974) for whom 'the general tendency of human evolution . . . consists in a growing emancipation from the forces of nature', which is accompanied by 'the emancipation from the domination of privileged individuals or groups'. *Emancipation*, considered as man's capacity to forge his own destiny and to realise fully his own potential, can be considered an end in itself, and the process of *development* as a gradual advancement towards this end.[3]

In order to activate a process of development which would approach emancipation progressively, it is necessary to consider that modern science has been demonstrated to be the most efficient means of generating knowledge for understanding the phenomena that surround man and, as Bronowski (1965) said, paraphrasing Bacon, of dominating not through force but rather through understanding. Moreover the technologies that emerge through systematic reflection (*logos*) about the repertoire of responses and practices that are available to act upon the physical and social world (*techne*), bestow an enormous power of manipulation to confront the challenges posed by the environment. Finally, productive and service activities associated with modern technology have acquired a huge potential to satisfy human needs. In this way, for a particular social group, and for those individuals that conform it, it is impossible to conceive an advance towards emancipation without a minimum level of autonomous capabilities to generate or adapt scientific knowledge, to transform it into technology, and to incorporate this technology, linked to scientific discoveries, into productive and service activities. This capacity has been called an *endogenous scientific and technological* base, and to take it into account becomes an indispensable requisite for the process of development (Sagasti, 1977).

Therefore the elements or components of the proposed conceptual framework can be summarised as follows: three currents of human activities (evolution of speculative thought, transformation of the technological base, and modification of productive and service activities); the social, cultural, and political context in which these three currents unfold; the interactions among these three currents, between them and their context, and between these currents and their counterparts in other societies; a global directionality for the evolution of these currents, contexts and interactions (the concepts of emancipation and development); and an instrumental condition (to take into account an endogenous scientific and technological base).

The unfolding and deployment of these components and the concrete forms they assume over time characterise the historical development of each society, and will also condition their future possibilities and options. An appreciation of the paths that have been covered in the past, conceptualised in terms of the proposed framework, would explain the present situation of backwardness of developing countries and make it possible to design strategies for overcoming it.

IV TOWARDS AN EXPLANATORY SCHEME

The proposed conceptual framework allows the development of an explanatory scheme that can throw light on the nature and present manifestations of the phenomena of development and underdevelopment. It is necessary to begin by recognising that in every society each of the three currents mentioned above, their contexts, and their interactions undergo a series of transformations in time. Nevertheless, considering a long historical period, the main transformations experienced by societies take place when there are major qualitative changes in the nature of speculative thought and in the process of knowledge generation. As a result of these changes the conceptions of man, about himself and about his relation to the physical world, will also evolve and expand progressively to encompass the technological base and the structure of produtive activities. Awarding the character of *primus inter pares* to the changes in the nature of speculative thought implies giving cognitive activities the role of primary ordering element in the explanatory scheme.

The challenge of the West

The evolution of the different societies can be examined in a relatively independent way until the period of the fifteenth to seventeenth centuries, during which the knowledge generation process underwent a radical transformation. Before this period, it is possible to analyse each society considered as an individual unit. Thus it is possible to examine, within reasonable limits, the European, Andean, Mayan, Aztec, Islamic, Chinese, and other cultures, employing the conceptual framework proposed here to follow the way in which the generation of knowledge, the technological base, and the productive activities evolved through history, and related to each other and to the wider context of social patterns, cultural activities and political processes.

However, the world suffered an irreversible transformation beginning with the scientific, bourgeois and industrial revolutions, which were accompanied by qualitative changes in the technological base and by the expansion of the productive capitalist system of Western Europe on a global scale. After that it is not possible to consider the evolution of societies in an independent way, and their study should take into account the challenges presented by the West to non-European society, as well as the responses that the latter generate. The point of inflection can be identified with the transformation of speculative thought and with the changes that took place in the generation of knowledge as a consequence of the scientific revolution. The transition towards a scientific conception of the world, through which it is possible to link systematically the abstractions and experiments on natural phenomena, to discover laws that rule the physical world, and to derive postulates, norms for action, and prescriptions that increase the domination of man over nature, constitutes an irreversible change in the evolution of humanity.

In parallel with these conceptual changes, and frequently associated with them, there were changes in the technological base, but these transformations were slower and would only accelerate two centuries later, when the number of production techniques based on scientific discoveries increased significantly. Notwithstanding their diffusion throughout all the regions of the world, beginning in the sixteenth century, the transformations in the nature of productive and service activities were even slower, although their pace has accelerated during the last century and a half, as a consequence of the ever closer relationship with technologies based on scientific discoveries.

The evolution of speculative thought

Examining briefly the evolution of speculative thought, it is clear that every culture presents its own way of generating and acquiring knowledge, but in general a transition can be observed from the contemplation and passive acceptance of nature towards a greater interaction between man and the phenomena that surround him. Whatever the scheme employed to explain this process – for example, Frazer's ideas (1964) on the transition from magic to religion and to science, or the alternative view provided by Malinowski (1974) – it is possible to perceive a progression towards the use of reason as the principal means to structure the human vision of the physical, social, intellectual, and even spiritual world.

However, to accept this progression and to recognise that the emergence of modern science was a point of inflection, does not necessarily imply the establishment of a great 'divide' between traditional pre-scientific knowledge and modern scientific knowledge. The wide spectrum of forms of knowledge generation must be considered as a continuum, without postulating a radical division between Western scientific thought and non-Western modes of thought. Elkana (1977) has examined this theme in detail.

> The tendency to link observed events by referring to theoretical entities, i.e., to make causal explanations, is a universal feature of human thought. Western science is distinctive, however, in creating deliberately new experiences by inventing theoretical entities in advance of common-sense observations – for example, non-Euclidean geometries (p. 160).

> There is in short no 'great divide' between Western science and traditional thought. There are no fundamental characteristics of the one which are totally absent in the other, no sources of knowledge unknown to either, no aims of knowledge acceptable to one of them only. (p. 161).

In this way, the changes in speculative thought and in the way of generating knowledge in different socieites present certain commonalities, although there are great variations in approach, rate of advance, and emphasis (for example, relative weight of abstract theories versus empirical aspects). This leads to a revaluation of the 'traditional' ways of generating knowledge that should be seen from a wider perspective, and not simply in comparison with the rigid and Europocentric pattern of Western science.

Furthermore the limits of Western science have begun to be emphasised recently and there have been suggestions that Western science is likely to suffer radical transformations in the near future.

Thompson (1978) argues that the descriptive ideal and concern of science will lead inexorably to a new type of mysticism and that physics has already been transformed from a materialistic science towards the processing of cognitive models. Thompson visualises a process of 'remythologization' and 'resacralization' in the field of human thought through science, art and religion that will converge in new modes of knowledge generation. Berman (1981) argues that it is necessary to develop a new 'enchanted' vision of the world that would incorporate the achievements of modern science, but that would also give man a sense of continuity in human experience and physical integrity, along with a sense of actively belonging in a cosmic scheme. De Riencourt (1981) proposes a synthesis of oriental mysticism and Western science, a theme treated earlier by Siu (1957) in his essay on Western knowledge and oriental wisdom. Snyder (1978) considers that the different cultural perspectives will enrich modern science, contributing to its future development. In any case, it is beyond doubt that the revaluation of non-Western forms of thought would contribute significantly to a better understanding of the transformations in speculative thought and more appropriate attitudes to the world that is emerging at present.

Changes in the technological base

When examining the evolution of the technological base in different societies, it is possible to appreciate that each of them has its own set of responses (techniques) to relate with the physical environment, and that this set will evolve gradually over time. In general, it is possible to postulate the transition from 'technical' stages towards more 'technological' stages. It could be said that, initially, a social group has at its disposal a set of 'techniques without technology' which encompasses a layer of passive empirical knowledge that only offers responses to specific challenges and situations; later it acquires a base of empirical knowledge that begins to detect variations and to register them through trial and error; and finally it develops a base of active empirical knowledge in which there is the beginning of systematic experimentation, but without theoretical constructs to orient the experiments. When advancing in this process of transition towards more complex and rich sets of techniques, the variety of available responses increases continuously until it constitutes a vast 'genetic reservoir' of techniques.

A subsequent stage is characterised by the evolution of technical responses as a function of theoretical constructions, thus moving from 'technique' to 'technology'. Initially the theoretical abstractions and reflections that support the advance of techniques are rudimentary and their impact is not very different from that of an active and systematised base of empirical knowledge. At this juncture it would be possible to speak about a 'incipient technology' or about a 'technological common sense'. Beyond this stage there emerge conceptions that explain the techniques and anticipate them, strengthening this transition towards technology. Pacey (1976) has characterised this transition as the move from the artisan to the technologist:

> The great strength of the technologist's discipline as compared with the craftman's art is that it allows him to design things by drawing and calculation which are outside the range of previous experience; it allows him to explore possibilities which are far beyond the point where the intuition of a practical man can offer any guidance. (p. 19)

Finally, and especially in the Western world, we arrive at the conception in which theory dominates technique, first through engineering and its professionalisation, and later through the almost direct incorporation of scientific discoveries into the development of new technologies. The triumph of 'technology' over 'technique' is now complete.

In this movement towards the predominance of the *'logos'* over *'techne'*, the variety of technical responses increases noticeably, but it acquires a potential rather than concrete character. That is to say, a large number of possible technological responses do not become a reality due, among other factors, to the fact that the conceptual schemes and theories that lead to their generation also contain criteria such as 'efficiency', 'reliability', 'simplicity' and other similar concepts that act as 'variety attenuators', limiting the process of transition from what is imagined theoretically to what is realised in practice. This leads to the apparent paradox of a greater variety or heterogeneity of effective technical responses in societies that have not been completely 'technologised', in comparison with those highly 'technologised' societies which present a relative homogeneity in their observable technical responses.

The challenges posed by the physical environment to a society, and the forms of organisation adopted to confront them, condition the demand for technical responses and the transformation of its technical base into a technological one, a process that also requires the development of a certain level of knowledge generation capabilities. As Needham (1977), Bernal (1971) and Alvarez (1979) indicate in their analysis of the technological achievements of non-Western cultures and societies, these acquired a set of technical and technological responses of their own, appropriate to their context, and processed by the forms of social organisation prevailing at the time. Therefore, now that the predominant forms of Western technological responses are being questioned,[4] it becomes important to study the alternative configurations of the technological base in societies that have not as yet been completely westernised.

Transformations in productive and service activities

The evolution of productive and service activities has as its principal motive the satisfaction of needs of the members of a social group, and is intimately related to the evolution of the processes of accumulation and the way in which the economic surplus is appropriated, distributed and allocated. However, the definition of 'needs' varies with time, with the degree of material development of a society and its income distribution patterns, and at present a large number of needs are generated artificially by the logic of the accumulation process itself, particularly in highly industrialised market-oriented economies.

An important turning-point in the evolution of productive activities is the new character that the process of accumulation assumed in the industrial civilisation of the West beginning with the bourgeois revolution. As Furtado (1979) has indicated, the reorientation of the process of accumulation towards the productive system, in such a way that surpluses are invested in the expansion of productive activities with the purpose of generating more surplus, conditions the evolution of the productive system and influences the transformation of the technological base. According to Furtado:

> In contrast with what takes place in traditional accumulation (in defence walls, in temples, in palaces. . .) that which is made in the productive forces seeks to obtain a surplus. This may come from the opening of new commercial routes, from the discovery of new natural resources, or from increases in the physical productivity of labour. This last case reflects the introduction of more efficient

methods which, in turn, are linked to a better division of labour or the use of better instruments. (p. 53).

These changes in the social organisation of production, which are a consequence of the way the surplus is used and the direction of the accumulation process, interact mutually with the transformations of the technological base and the evolution of speculative thought. The expanded repertoire of technological responses presents the productive system with a range of possibilities for increasing the generation of surplus, while the greater surplus available constitutes a challenge to human inventiveness and stimulates the development of new technologies. On the other hand, the emergence of the secular concept of reason, the desacralisation of nature, and the rational conception of the world that finds its expression in thinkers like Descartes and Bacon, gave ideological support to the organisation of production in accordance with the demands of the process of accumulation, and also with the appropriation of the surplus associated with the emergence of capitalism. At the same time, the diffusion of capitalist production, characteristic of the industrial civilisation of the West, contributed to the predominance of the secularised and instrumentalist vision of 'rationality' which expanded its scope progressively, even reaching the very conception of human relations (Barret, 1979; Berger *et al.*, 1974).

A constant in the process of evolution of productive activities, particularly during the last four centuries, with the diffusion of capitalism and the industrial civilisation of the West, has been the enlargement of their geographical scope. From their organisation at the local level, production and service activities extended at regional and continental levels, and at present encompass the whole planet. This internationalisation process has been accompanied by the emergence of a global consumer elite with relatively uniform consumption patterns, superimposed upon a variety of local forms of consumption – corresponding to much lower levels of income and resource use – in the underdeveloped societies.

A central issue in the discussion of the development problematique refers to the paths that countries of the Third World, which lack the capacity for accumulation of the Western industrialised nations, should follow in order to expand their productive and service activities and satisfy the needs of the population. While it is clear that the process of development, whatever its conception, entails the satisfaction of material needs at a level compatible with human dignity, the evolution of productive and service activities need not necessarily follow the same path as that followed by the highly industrialised nations, particularly with regard to the volume and diversification of goods. Concepts such as 'another development' (see, for example, the Dag Hammarskjold Report, 1975) question this premise and seek to propose different options that would involve a redefinition of needs, and even the revaluation of productive and service activities of societies outside the European–American sphere.

Interactions among the three currents

The interactions among the different stages in the evolution of the three currents, visualised against the background of the social, political and cultural organisation, would characterise the degree of development of a given society. For example, in the West the evolution of speculative thought led to science as the key method for generating knowledge, which accelerated the transformation of the technological base and helped in the transition from 'technique' to 'technology' while receiving at the same time the support of many technological advances which contributed to the scientific enterprise. Productive and service activities found increasing support in the new technologies related to science, to the extent that at present productive activities which employ technologies of scientific origin are clearly superior and dominate the scene. All of this takes place simultaneously with the acceleration and reorientation of the process of accumulation and with the emergence and expansion of capitalism as the dominant mode of production, a process which feeds on the technological and scientific advances and which, in turn, gives the stimulus and the material resources to support them. This process has been called the rise of an *endogenous scientific and technological base* in the highly industrialised countries.

The emergence of an endogenous scientific and technological base is accompanied by changes in values, with a new vision of the physical, social and intellectual world, and with a series of changes related to the diffusion of the by-products of the scientific activities. All of this gives its specificity to Western culture. In parallel, other cultures and societies have developed their own ways of linking the three currents and of relating them to the social, political and cultural context. For example, China displayed great achievements in the evolution of speculative thought about nature, as well as in logic and mathematics; she was able to generate technologies based on abstract and systematic conceptions, and developed an efficient social organisation of production. As Needham has pointed out (1977, p. 194), the philosophical and intellectual tradition of China at the time of the Renaissance was 'much more congruent with modern science than the Christian conception of the world.'

However, a variety of social, economic, and political factors – which emerged as a response to the specific environment of Chinese culture – did not conduce to modern science and to an endogenous scientific and technological base. Similar considerations can be applied to India, the Islamic world, and to cultures of other regions. By examining their transformation it would be possible to identify the variants or the different 'models' of societal development, without falling into a spurious comparison with the achievements of Western civilisation taken as a frame of reference. In this way it would be possible to develop a proper perspective for examining the achievements of the West, for understanding its limitations and the nature of its present crisis, and for exploring new roads to the progressive acquisition of an endogenous scientific and technological base in Third World countries.

Towards a new cultural context

Seen in this light, the progressive establishment of an endogenous scientific and technological base in the non-Western countries requires a new cultural context, different from the one that characterised the emergence of industrial civilisation. For this it would be necessary to transcend the narrow or rationalistic vision characteristic of that civilisation, avoiding the almost complete subordination of creativity and of knowledge generation to the logic of the productive process. The new cultural context must leave ample room for the idea of emancipation as the directionality for the evolution of human groups, should accommodate the diversity of the products of social inventiveness, and should also restructure the pattern of values of industrial civilisation which privileges the means, and embraces instrumental rationality (emphasis on 'how'), rather than the ends, and the conception of a human destiny (emphasis on 'why'). The idea is to overcome the situation that Furtado has described in the following terms:

> The most fundamental impulses of man, generated by the need for self-identification and for defining his place in the universe –

impulses that constitute the matrix of creative activities: philosophical reflexion, mystical meditation, artistic invention, and basic
scientific research – were subordinated, in one way or another, to
the process of transformation of the physical world required by the
drive towards accumulation. This led to the atrophy of the linkages
between creativity and human life conceived as an end in itself,
and to the hypertrophia of the links between creativity and the
instruments used by man to transform the world. (Furtado, 1979,
p. 100).

The new cultural context required for the establishment of an
endogenous scientific and technological base in the Third World
would seek to rescue creativity from its subordination to the productive process, to diminish the importance of instrumental rationality,
and to counteract the homogenising tendencies associated with
industrial civilisation. In this way, taking into account the reassertion
of a diversity of finalities, which is the counterpart of these changes,
there will be room for a greater variety of ways of articulating
speculative thought with regard to the technological base and productive activities. This perspective has been highlighted by Ladriere
(1977):

The type of culture that appears to announce itself through the
interactions between the technical and scientific systems, and the
different cultural subsystems, is a culture crossed by multiple
tensions, which suggest diverse modes of articulation among its
own components, and also between the other systems and itself,
which proposes a variety of schemes for action and is flexible in its
own structure, compatible with multiple forms of equilibrium. . . ..
In that culture there is not, properly speaking, a unique center that
integrates everything, but multiple centers. A relative dispersion
substitutes unity. (p. 193).

In another essay I have speculated about the future of present day
societies, postulating that we are witnessing the appearance of
two 'civilisations'. The first civilisation corresponds to the highly
industrialised countries that have an endogenous scientific and technological base, and the second civilisation to those Third World countries
that lack it. To combat the threat posed by a growing division of the
human race into two distinct and antagonistic camps, it will be
necessary to advance towards a 'third civilisation' in which the

achievements of modern science could be integrated in a harmonious fashion with the cultural heritage of non-Western societies. This search for a third civilisation, which must be considered as a general frame of reference within each society could explore its own paths, requires many conceptual changes and a refocussing of our perception of Western culture:

> Despite its unquestionable achievements, the Western scientific–technological culture of the first civilization should not be considered as the universal model to be imitated by the countries of the second civilization; it should rather be viewed as one of the many phases of a global and historial process of material and intellectual evolution. There is a need to discard the implicit arrogance of Western culture which makes the first civilization consider itself as the model to be followed by the developing world. A more ecumenical perception of the processes of development and progress is required, in which the potentialities of the many cultures that are part of the second civilization would be revalued and appreciated, particularly if we have the foresight to visualize what could be achieved if a harmonious integration of their cultural heritage with modern science were possible. (Sagasti, 1980, p. 132).

Notes

1. For a review of development theory see Björn Hettne, *Development Theory and the Third World*, Stockholm, SAREC Report 22: 1982; the treatment of technology in several theories is discussed in my monograph, *A Review of Schools of Thought on Science, Technology, Development and Technical Change* (STPI Module 1), Ottawa, International Development Research Centre, 1980.
2. The project ' A Scientific and Technological Reinterpretation of Development' is being carried out at the Group of Analysis for Development (GRADE). It received support from the Swedish Agency for Research Co-operation with Developing Countries (SAREC) during its first stage.
3. I am conscious that this involves bringing an exogenous, pre-defined, concept into the model, that of 'emancipation', but short of falling into an absolute cultural relativism, or giving a particular culture the status of a frame of reference, I see no other way of providing a framework of comparison.
4. See, among others, the works of Ellul (1980), Winner (1977), Schwartz (1971) and Schumacher (1973).

References

Alvarez, Claude (1979) *Homo Faber: Technology and Culture in India, China and the West, 1500–1972* (New Delhi: Allied Publishers Private Limited).

Barret, William (1979) *The Illusion of Technique* (New York: Doubleday).

Basalla, George (1967) 'The Spread of Western Science', *Science*, vol. 156 (May, 1967) pp. 611–22.

Berger, Peter, Brigitte Berger and Hansfield Keller (1974) *The Homeless Mind* (Harmondsworth: Penguin).

Berman, Morris (1981) *The Reenchantment of the World* (Ithaca: Cornell University Press).

Bernal, J. D. (1971) *Science in History* (Cambridge, Mass: The Massachusetts Institute of Technology Press).

Braudel, Fernand (1975) *Las Civilizaciones Actuales* (Madrid: Editorial Tecnos).

Bronowski, Jacob (1965) *Science and Human Values* (New York: Harper Torchbooks).

de Riencourt, Amaury (1981) *The Eye of Shiva: Eastern Mysticism and Science* (New York: Morrow Quill Paperbacks).

Elkana, Yehuda (1977) 'The distinctiveness and universality of Science. Reflections on the work of Professor Robin Horton', *Minerva*, vol. 15, no. 2 (Summer 1977) pp. 155–73.

Ellul, Jacques (1980) *The Technological System* (New York: The Continuum Publishing Corporation).

Frazer, Sir James (1964) *The New Golden Bough* (New York: Mentor Books).

Furtado, Celso (1962) *Desarrollo y Subdesarrollo* (Buenos Aires: EUDEBA).

Furtado, Celso (1979) *Creatividad y Dependencia* (Mexico D.F.: Siglo XXI Editores).

Garaudy, Roger (1981) Notes for a Keynote Search at the Fifth Parliamentary and Scientific Conference of the Council of Europe, Helsinki, June 1981.

Informe Dag Hammarskjold (1975) *Qué Hacer?* (Número especial de *Development Dialogue* (Uppsala: Fundación Dag Hammarskjold).

Jaguaribe, Helio 'Ciencia y tecnología en el cuadro socio-político de América Latina', *El Trimestre Económico*, No. 150, Abril–Junio 1971, pp. 389–432.

Ladriere, Jean (1977) *El Reto de la Racionalidad: la Ciencia la Tecnología frente a las culturas* (Madrid: UNESCO/Ediciones Sígueme).

Malinowski, Bronislaw (1974) *Magia, Ciencia, Religión* (Barcelona: Editorial Ariel S.A.).

Mumford, Lewis (1970) *The Pentagon of Power: The Myth of the Machine* (New York: Harcourt Brace, Jovanovich).

Mumford, Lewis (1971) *Técnica y Civilización* (Madrid: Alianza Editorial).

Murra, J. (1975) *Formaciones Económicas y Políticas el Mundo Andino* (Lima: Instituto de Estudios Peruanos).

Nasr, Seyyed Hossein (1970) *Science and Civilization in Islam* (New York: Plume Books).

Needham, Joseph (1977) *La gran Titulación: Ciencia y Sociedad en Oriente y Occidente* (Madrid: Alianza Editorial).

Nisbet, Robert (1980) *History of the Idea of Progress* (New York: Basic Books).

Ortega y Gasset, José (1968) *Meditación de la Técnica* (Madrid: El Arquero, Revista de Occidente).

Pacey, Arnold (1976) *The Maze of Ingenuity: Ideas and Idealism in the Development of Technology* (Cambridge, Mass: The Massachusetts Institute of Technology).

Patiño, Víctor Manuel (1965) *Historia de la Actividad Agropecuaria en América Equinoccial* (Cali: Imprenta Departamental).

Prebisch, Raul (1952) 'Problemas Teóricos y Prácticos del crecimiento económico', New York: Comisión Económica para América Latina.

Ravines, Rogger (Compilados, 1978) *Tecnología Andina* (Lima: Instituto de Estudios Peruanos).

Roche, Marcel (1976) 'Early History of Science in Spanish America', *Science*, vol. 194, pp. 806–10.

Sagasti, Francisco (1976) 'Algunas ideas para una estrategia de desarrollo científico y tecnológico', *Cuadernos del Consejo Nacional de la Universidad Peruana*, nos. 22–23, Julio–Diciembre.

Sagasti, Francisco (1977) 'Reflexiones sobre la Endogenización de la Revolución Científico–Tecnológica en Países Subdesarrollados', *Interciencia*, vol. 2, no. 4, Julio–Agosto, pp. 216–21).

Sagasti, Francisco (1978a) 'Ezbozo Histórico de la Ciencia y la Tecnología en América Latina', *Interciencia*, vol. 3, no. 6, Noviembre–Diciembre, pp. 351–9.

Sagasti, Francisco (1978b) *Ciencia y Tecnología para el Desarrollo: Informe Comparativo del Proyecto STPI* (Bogotá: Centro Internacional de Investigaciones para el Desarrollo).

Sagasti, Francisco (1979) 'Towards Endogenous Science and Technology for another Development', *Development Dialogue*, no. 1, pp. 13–23.

Sagasti, Francisco (1980) 'The two civilizations and the process of development', *Prospects*, vol. X, no. 2, pp. 123–39.

Sardar, Ziauddin (1977) *Science, Technology and Development in the Muslim World* (London: Croom Helm).

Schumacher, E. F. (1973) *Small is Beautiful: Economics as if People Mattered* (New York: Harper & Row).

Schwartz, Eugene S. (1971) *Overskill: The Decline of Technology in Modern Civilization* (Chicago: Quadrangle Books).

Siu, R. G. H. (1957) *The Tao of Science* (Cambridge, Mass: The Massachusetts Institute of Technology Press).

Snyder, Paul (1978) *Towards One Science: the convergence of traditions* (New York: St Martin's Press).

Thompson, William Irvin (1978) *Darkness and Scattered Light* (New York: Anchor Press, Doubleday).

Weinberg, Gregorio (s.f.) 'Consideraciones sobre la historia de la tradición científica en el desarrollo de la conciencia social y su importancia en la formación de la conciencia nacional y latinoamericana', Buenos Aires, mimeo.

Wertheim, W. F. (1974) *Evolution and Revolution: The Rising Waves of Emancipation* (Harmondsworth: Penguin).
Winner, Langdon (1977) *Autonomous Technology* (Cambridge, Mass: The Massachusetts Institute of Technology).

4 Socially Viable Ideas of Nature: a Cultural Hypothesis

Michael Thompson

I CONTRADICTORY CERTAINTIES AND THE DISCERNING SPECTATOR

What is a resource? I was once fortunate enough to be spectator to an exchange of views on this question between a distinguished ecologist and a Nobel Prize-winning physicist. The ecologist let drop something about 'natural resources' and the physicist was down on him like a ton of bricks. 'You cannot talk about *natural resources*,' he cried, 'there are only *raw materials*', and he went on to explain how a raw material only becomes a resource when human ingenuity, skill and enterprise are successfully focussed upon it. This is a profound and insoluble disagreement. For our ecologist riches are given to us by nature; for our physicist[1] they are given to us by our social inheritance – by that complex whole that gets transferred from one generation to the next by mechanisms that are not genetic; a whole that includes the whole of language, the whole of knowledge, the whole of technology, and a great deal more besides.

Clearly our ecologist and our physicist locate resources very differently. Their premises, in other words, are different and, as a result, so are the sorts of policies that they see as desirable (or even feasible). Our ecologist has an *idea of nature* as something stern and unforgiving – as supplying him with a strictly *accountable* inventory of resources. Our physicist, on the other hand, sees these limitations as being of little consequence because they are capable of modification, exploitation and multiplication through the application of skills that are socially acquired and transmitted. In this way he is led to the idea of nature as essentially *cornucopian*. So here is a fundamental cleavage. Our physicist's world is a world of *resource abundance*; our ecologist's one of *resource depletion*.

When people argue from different premises they will, in all probability, fail to agree. At best, they may agree to differ. This is

57

something of a disappointment to those bystanders who want to know *the* answer, but the discerning spectator (the individual whose viewpoint I am urging we should adopt) is not one of those. His attention is focussed not on the facts of the matter but on the facts of the disagreement. He is not looking for the single truth 'out there' but at the various convictions 'in here'. For him what is being argued about is a foreground distraction and, disregarding this, he zeroes in on what really interests him: the premises and their differences. 'Where do these premises come from?' 'How many kinds of premises are possible?' 'What leads this individual to this premise and that individual to that premise?' These are the sorts of questions that the discerning spectator asks, not 'Who is right?' In other words, the discerning spectator begins by granting legitimacy to all these sets of contradictory premises. Nor does the fact that they are contradictory cause him any dismay. On the contrary, he sees social life as a process that depends for its very existence on the perpetual contention between these different sets of convictions about how the world is.

II   THE CULTURAL HYPOTHESIS

When we look at our environment we do not see it with the naked eye. We see it as it is filtered through a cultural screen – our idea of nature. From this there follows a very minimal definition of human rationality;[2] an act is rational if it is consistent with the actor's idea of nature. The only trouble with this definition is that it would seem to insist that *every* act is rational. No matter how bizarre an act may be we have only to dream up and ascribe to the actor a correspondingly bizarre set of beliefs about how the world is for that act to become rational.[3] But this descent into complete relativism would be inevitable only if, first, there were no social constraints on the beliefs people could adhere to and, second, there were no chance of nature herself pointing out, from time to time, the inadequacies of some of those beliefs. If we allow for the existence of these two sorts of constraints then, instead of complete relativism, we obtain *a system of constrained relativism* – a system that enables us to avoid the dissipative nihilism of the relativist position without at the same time succumbing to the narrow tyranny of the universalist position.

The social constraints

The cultural hypothesis holds that beliefs, though varied, are not free to float about just anywhere. They are closely tied to the social situations that they help sustain and render meaningful. If the number of kinds of social situations that are possible is limited (and the hypothesis claims that it is) then so too will the variation of belief be limited to a quite small number of distinctive patterns or *cultural biases*. If ideas of nature have to pass the test of social viability then the weeding out of the non-viable ideas will leave us with a multiplicity, but not an infinitude, of mutually contradictory ideas each one of which will be associated, through a process of mutual reinforcement, with a distinctive patterning of social relations.

The whole retreat from rationality, which might appear to be relativism's inevitable accompaniment, can be averted if we say that metaphysical beliefs are embedded in culture, and that culture is not some dead weight of habits that is passed on unchanged from generation to generation, but a lively and responsive thing that is continually being negotiated and renegotiated in order to sustain and justify preferred patterns of social relationships. It is the *adaptive* propensity of this negotiating process, coupled with the *adoptive* criteria of the social environment in which it takes place, that gives rise to the first dynamic component of the system – the cultural construction of nature.

The natural constraints

That Doctor Johnson was able to refute Bishop Berkeley by kicking a stone should remind us that nature herself can sometimes provide negative feedback to curb the wilder excesses of the relativistic urge. But this sort of feedback does not always get through. In a social setting where everyone subscribes to the same idea of nature there will be no sceptic around to deliberately kick the stone. Of course, if the natural constraint is there, people will sometimes kick stones by accident but it is wonderful what we can collectively manage not to see.

An idea of nature furnishes us with a way of seeing the world and, more importantly, with a way of not seeing the world – it actually filters out most of the negative feedback.[4] This means that, only when the cumulative costs of maintaining that idea have built up into an intolerable burden, will the negative feedback finally force its way through the cultural filter and be noticed. When this happens we suffer *surprise*. In much the same way that the only lasting laws of nature are negative (cannot do) laws, so no event is absolutely surprising. It is only surprising if, first, it is contradictory in relation to a particular idea of nature and, second, it is noticed. Then, and only then, will a socially desirable element of belief come into direct conflict with an implacable nature. This – the natural destruction of culture – is the second dynamic component that completes the system.

So the full theory of constrained relativism has to take account of both sets of constraints, and of their interaction. It has to synthesise the theory of plural rationality (which has been developed largely by anthropologists) and the theory of surprise (which has been developed largely by ecologists). But, since that synthesis requires some lengthy and complex argumentation, I will not present it here.[5] Here I will restrict myself to exploring the plural rationalities and to drawing out their implications for policy analysis.

The retreat from rationality

Let me begin by contrasting the notions of rationality that are built into two approaches to policy analysis: *classic decision theory* and what, for want of a better label, I will call *historical contingency theory*. I take as an exemplar of the first approach Myron Tribus's standard text *Rational descriptions, decisions and designs*.[6]

> To develop criteria for . . . decisions we need to define what we mean by *rational*. We shall say that a person who knowingly makes a decision which is against his own stated objectives is behaving irrationally. That is, if a man asserts that he wishes to accomplish an action, say A, and he deliberately takes action B which he knows will thwart action A, then to the extent that he told us the truth about A, he is acting irrationally. We shall not, in this book, consider what to do about whether or not the man told the truth about A. That is a task for the psychiatrist. Rather we shall take the stated goal, A, as correct and develop aids to help decide if the actions are consistent with A. (Tribus, 1969).

The policy analyst might well find this restriction to just the stated goal rather unhelpful. What he would like is some approach that would enable him to handle the hidden agendas (as he calls them) as well as the visible ones. And, given his daily familiarity with hidden agendas, he may feel that it is he, not the psychiatrist, who is best equipped to understand the sort of strategising behaviour that sometimes leads a person to take action B when his stated aim is action A. The policy analyst (when he can advance some plausible hypothesis in terms of a hidden agenda and an appropriate strategy for its advancement) would dearly like to be able to extend rationality to these sorts of actions.

So classic decision theory certainly cannot handle everything; but, surely, within its explicit and self-imposed limits it is valid and useful. Historical contingency theory takes issue even with this seemingly modest and innocuous claim. Classic decision theory, it concedes, can of course claim validity in all those cases where people know (and say) what their objective is but historical contingency theory holds that this is a category with no members.

> Nobody knows what their 'real' interests are. It is a very fundamental principle that nobody knows what is the ultimate effect of almost any act whatsoever . . . The actual consequences of almost any act are unknown and unforeseen, which is a little rough on the theory of rational behaviour.[7] (Boulding, 1983).

Of course, Boulding may be wrong but, even if he is, classic decision theory would still only be a theory of *goal-seeking*; we would still be without any theory of *goal-setting*. And, if Boulding is right, then we are left without any theory at all . . . of goal-setting *or* goal-seeking.

Redefining the decision theorist's problem

Real policies, unlike the decisions analysed by decision theory, usually involve a variety of actors and interests, conflicting perceptions of nature, contradictory rationalities and divergent advocacies. They are not static phenomena but historical processes. The rapid development of decision theory, coupled with the decision-maker's desire to know which out of a bewildering array of counsels is the 'right' one, has propelled policy analysis towards the fallacy of misplaced concreteness – towards a pretence that things are tidier than they really are. But complexity, goal ambiguity, contradictory certainties, conflict, institutional inertia, and temporal change are not disfiguring warts on the face of policy; they are its essential characteristics. The central problem, therefore, is to resist the urge to remove them and to re-conceive policy in a way that preserves its historical contingency.

Only when this has been done can we adequately understand an evolving process and the extent to which we can both manage its evolution to fit our desires and adapt our desires to fit its evolution.

So how do we re-conceive policy – warts and all? Let me suggest that we put goal-seeking to one side for the moment and begin with the really big question: 'How do the goals that people seek get set?' And let me suggest that we approach it by looking at socially viable ideas of nature. Socially viable ideas of nature correspond to what Mary Douglas[8] has called *cultural biases* – those sets of shared beliefs and convictions about how the universe is that sustain and justify moral judgments.

III SOCIALLY VIABLE IDEAS OF NATURE

The cultural hypothesis holds that there are just five distinct cultural biases each of which has associated with it a distinct idea of nature. Each of these conjunctions of cultural bias and idea of nature finds itself adopted in one particular reach of social life and rejected in all of the others. The different reaches of social life are described by the two axes of *social context: group*, which has to do with the extent to which an individual is incorporated into or free from bounded social groups, and *grid*, which has to do with the extent to which he is subject to or free from socially-imposed prescriptions. Since the processes that give rise to (and sustain) group-formation and prescription-imposition are dynamic processes, group inclusion here implies group exclusion somewhere else and being subject to prescription here implies subjecting to prescription somewhere else. In other words, the group and grid axes have both positive and negative directions.

Since group and grid can only be measured on ordinal scales, there are only five distinctions to be made within this social context space – one at the origin and one in each of the four quandrants. In each of these distinct social contexts we find a distinct social being: at the centre, *the hermit*, free from coercive involvement in both group-formation and personal network-building; at the bottom left, *the entrepreneur*, spurning group involvement and central to a large personal network; at the top left, *the ineffectual*, excluded from social groups and peripheral to the personal networks of others; at top right, *the hierarchist*, strongly grouped and willingly subject to all the prescriptions that serve to maintain the ranked separation of his

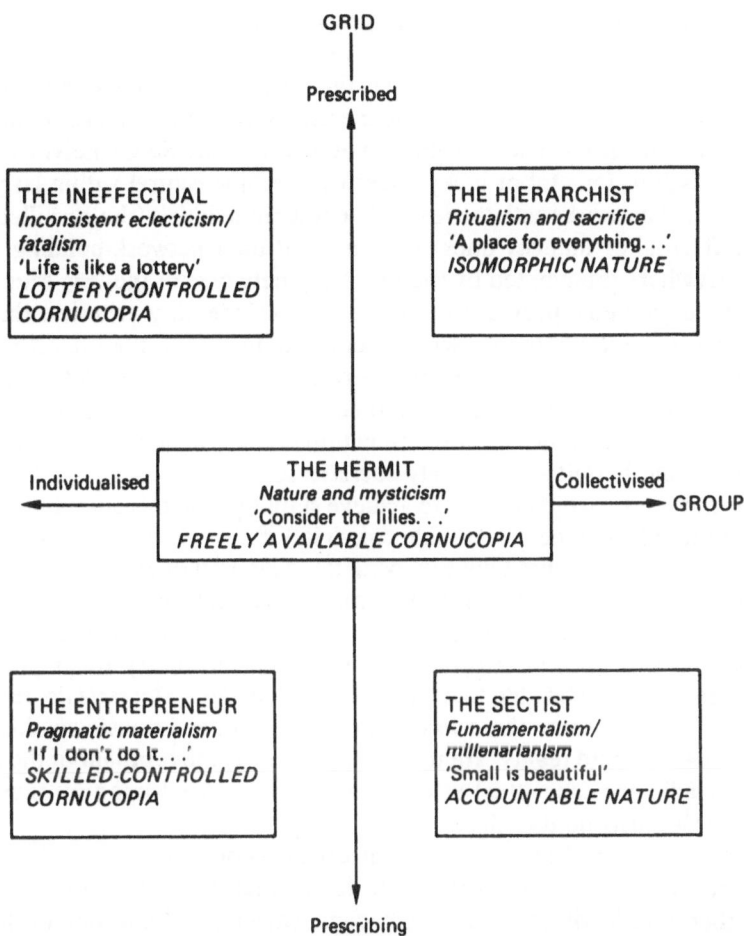

Figure 4.1 SOCIAL BEINGS, *cultural biases*, 'justifications' and IDEAS OF NATURE

group from all the others within the group hierarchy; and at bottom right, *the sectist*, strongly grouped but rejecting hierarchy and all the prescriptions that are its inevitable accompaniment (Figure 4.1).

I trace these five stabilisable conjunctions of social context and cultural bias back to three distinctive kinds of organisation: *the ego-focussed network, the hierarchically-nested group* and *the bounded egalitarian group*. I further argue that this typology of organisations is exhaustive – that these are the only kinds of organisation that are socially viable. But how can just three kinds of organisation give rise to five cultural biases? Such a proposition would appear to run counter to Ashby's law of requisite variety.[9]

The answer has two parts. First, the process of personal network-building will result (so long as there exist opportunities for economies of scale) in an asymmetrical pattern of involvement. Some individuals (the skilful, forceful or lucky ones) will become central within large personal networks with the result that other individuals (the unskilled, ineffectual or unlucky ones) will find that their network-building is everywhere pre-empted by the ramifying networks of the 'Big Men'. This asymmetry provides the centrality/peripherality criterion that serves to separate the prescribing entrepreneur from the prescribed ineffectual. Second, a personal strategy aimed at the deliberate avoidance of all three organisational forms can also (under certain conditions) result in a viable conjunction of social context and cultural bias – the hermit's. (It is only *coercive* social involvement that the hermit has to avoid in order to achieve stability, and this autonomous cultural bias can be remarkable convivial).[10]

I should stress that I am following the definition of an organisation as a *conceptual scheme*. I do not wish to suggest that the 'concrete reality' – the process of social life – crystallises out so neatly. In general this process is sufficiently complex and messy for anyone involved in it to be able to conceive it, *and render a plausible account of it*, in one of these three ways. The patternings and transformations of this 'concrete reality' are to be understood as the resultant of these contradictory conceptual schemes as they are acted upon by those who variously hold to them.

So the central idea is that each organisational form has, all the time, to generate within itself the forces that will hold it together. Otherwise, it will fall apart (or become transformed into one of the others). Ideas of nature, and the moral justifications that they provide the basis for, are the means by which these organisational needs are met. In each context just one idea of nature is capable of providing the necessary stabilising forces and all the others would inevitably result in its transformation.

For example, pragmatic materialism provides a strategy (or behavioural programme) that reinforces the social context of economic individualism and, at the same time, receives its moral justification (and shareability – the necessary condition for moral community) from the 'skill-controlled cornucopia' idea of nature. The great moral justification for economic individualism is 'the hidden hand' that steadily adds to the welfare of the whole, and it would lose all its validity if life were revealed to be a zero-sum game (or, worse still, a negative-sum game). A strictly accountable nature, therefore, is unthinkable. Nature *must* be cornucopian. The skill-controlled part of this idea of nature provides the basis for the other great moral justification of economic individualism – equality of opportunity. If fortune favours the brave, if unused talents atrophy, if faint heart never wins fair lady, if there is a tide in the affairs of men that must be taken at the flood, if a fool and his money are soon parted . . . if nothing succeeds like success, then inequality of result can never be a moral reproach to those who have acted with skill and daring.

Conversely, those whose results are less impressive can fashion for themselves a *modus vivendi* by making just a small modification to this cornucopian idea of nature. Where the successful emphasise skill and daring, the unsuccessful can emphasise luck (and a measure of unfair advantage). If it's all in the stars, if your number is on it, if it's your (or, more likely, his) lucky day, if it's the same the whole world over . . . if it's always the rich what gets the gravy, then the erratic payouts and withholdings can all be understood in terms of a cornucopian nature that is controlled not by skill but by lottery – a one-armed bandit on the cosmic scale. When, as occasionally happens, the jackpot comes your way . . . oh happy day! When, as usually happens, it does not . . . those crafty bastards have got it fixed.

The hermit's social context, likewise, is individualised – there are no bounded groups around – but, unlike the contexts of the entrepreneur and the ineffectual, there are no economies of scale around either. Cosmopolitan Sherpas who can grow their own potatoes and raise their own yaks, international owner–driver haulage contractors who can only drive one lorry at a time . . . easy-going caretakers who can only look after one modest office building at a time, have little incentive to build and maintain vast coercive personal networks. What would they use them for? Their social isolation means that, if they are not economically viable, they will not be around for long and so this means that virtually all those hermits that are around are economically viable. If enough comes in (and the fact that they are still there means that enough is coming in) then enough is enough: if excessive effort just leads to heavy scenes, if pissing matches with skunks are always disappointing. . . . if in getting and spending we lay waste our powers, then what we need are some gentle moral marker flags to prevent us from inadvertently straying into coercive social involvement (be it in personal networks or group formation). Take, therefore, no thought for the morrow; consider, instead, the lilies of the field . . . get your autonomous act together, man, and nature will provide. For such a quietist morality to remain credible, nature must be cornucopian, but it cannot be skill-controlled nor can it be lottery-controlled. It must be freely available.

In this way one idea of nature – the cornucopian – sustains all three individualised contexts. But each context modifies that idea to suit its particular manifestation of the single organisational type that provides their common origin: the ego-focussed network. With no economies of scale networks remain little developed and the cornucopia remains freely available. As economies of scale are introduced so a number of things happen. Networks become competitively developed, an asymmetry opens up between those who are central and those who are peripheral, the cornucopia becomes controlled, and the mode of that control bifurcates between skill and lottery (Figure 4.2).[11]

Where the tragedy of competitive individualism is the tragedy of the commons, the tragedy of the bounded egalitarian group (the sect)

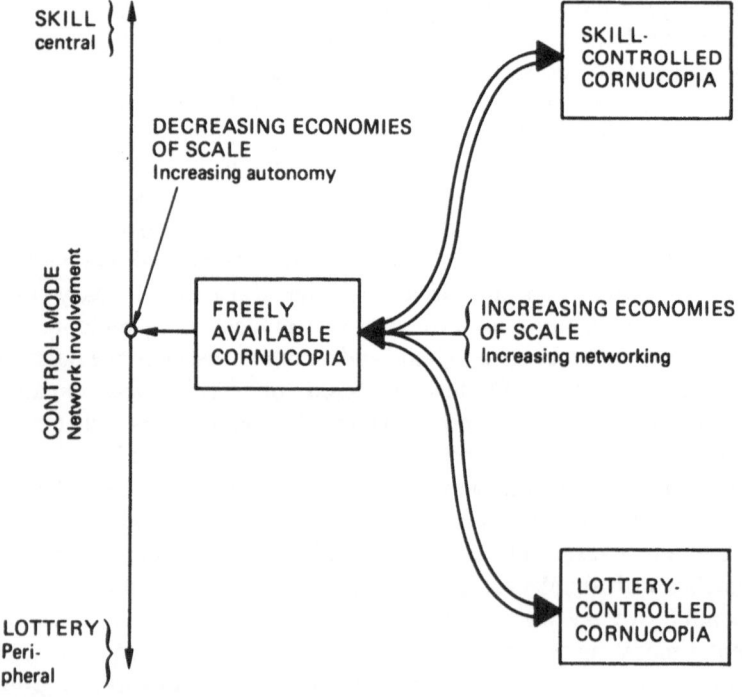

Figure 4.2   The three cornucopian ideas of nature

is almost the exact reverse – the tragedy of the crabs in a barrel. In the West Indies fishermen will put their day's catch of live crabs into a barrel. Though crabs are good climbers, the fishermen do not bother to put a lid on top of the barrel because no sooner does one crab climb up towards the rim than it is immediately pulled back down by its fellows. All the crabs could escape if only they were prepared to allow some to go first. But they are not, and they all perish.

The great moral justification needed to sustain the sect is equality, not of opportunity, but of result. To allow priority to some for the sake of the benefits that will eventually accrue to all would be to transgress the rule of absolute equality which, in the absence of the sorts of differentiations that exist in markets and in bureaucracies, has to be the sect's sole organising principle. Positive-sum games, in consequence, are unthinkable. Life has to be a zero-sum (or, better still, a negative-sum) game. A world of resource depletion is the environment best suited to the nurture of a bounded egalitarian group. And a world of resource depletion is guaranteed by an idea of nature as strictly accountable.

Accountable nature also provides the basis for the sect's other great moral justification – that which serves to maintain the sharpness of the boundary that separates the saved on the inside from the damned on the outside. The sect members, by their insistence on equality of result, make sure that they respect Nature's fragile limits. It is those in the wicked world beyond the sect that are misusing her. In this way the boundary between inside and outside is sharply drawn between those who respect Nature and those who abuse her. Blame – system blame – can then be exported.

Both the cornucopian and the accountable ideas of nature would wreak havoc inside that complex edifice, a hierarchical collectivity. The first would undermine the boundaries that sustain its highly discriminated structure; the second would erode the status differences that those boundaries uphold. But, of course, the sort of environment that is created by hierarchically-nested groups rejects both these ideas of nature. The mutual reinforcement that it needs to sustain its existence has no place for individual salvation and no place for equality of result. Rather, it is all geared up to adopt a collectivised and stratified mode of salvation – everyone in Peter's barque but with first-class, second-class and steerage passengers! In a hierarchy, all men do not end up equal; it is this that distinguishes it from a sect. Nor do they all start off equal; it is this that distinguishes it from an ego-focussed network. In consequence, neither the moral principle of equality of result nor that of equality of opportunity can mesh with the premise of inequality that sustains a hierarchy. Rather, a hierarchy will stress equality before the law – a hierarchical law that embodies the premise of inequality and entitles those of high rank to be tried by their peers. Peer review – the established method of assessment in the scientific community – provides a nice example of this moral principle at work.

So hierarchy needs rather complicated moral justifications if it is not to be eroded. It needs to justify inequality and it needs to justify separation. Cornucopian nature, with its positive-sum game, would justify inequality but would be destructive of separation. Accountable nature, on the other hand, would justify separation but would pillory inequality. The solution is an *isomorphic nature* that does permit positive-sum games but within certain defined bounds.

Seen thus nature and society are both complex, yet clearly separate, systems. Though separate they are isomorphic; nature, as it were, holds up a mirror to society. If nature is a positive-sum game (and there is nothing in this mirror idea to insist that it is not) then so too is society. If riches bubble up in nature, it is probably thanks to nature's clearly understood complexity; and, if they are bubbling up in society, it is probably because they are the accurate reflection of the positive-sum benefits that flow from the division of labour and status within society. But, if the complexity of the two matched systems is the source of these collective benefits, then it is absolutely vital that the clarity and resolution of the mirror be maintained – that the isomorphism be assiduously preserved. Clarity, predictability, discrimination, resolving power and order – these are the great moral imperatives that are generated by the isomorphic idea of nature. Look after them and the positive-sum benefits will look after themselves.

Political cultures and part-regimes

In tracing social contexts back to viable organisational forms, and in tracing cultural biases back to shareable ideas of nature, we are able to uncover a very general self-segregating system within which certain conjunctions of social environment and ideas of nature become (by a process of moral justification) mutually reinforcing while other conjunctions (by a process of moral indefensibility) become mutually repulsive. It is this system of attractive and repulsive forces that ever maintains the possibility for the existence of the five stabilisable conjunctions.

So these are the eternal bases, as it were, onto which we home in; and it is this homing-in process that induces and maintains the distinctive *personal strategies* that go with each cultural bias: two manipulative strategies – the individualist and the collectivist – that are adopted by the entrepreneur and the hierarchist respectively; two survival strategies – the individualist and the collectivist – that are adopted by the ineffectual and the sectist respectively; and one autonomous strategy that, by steering clear of the sorts of social involvement that inevitably result in manipulating others or in being manipulated, soon recommends itself to the hermit.

When these socially-induced personal strategies are combined with the historical processes of change, they become imbued with a sense of direction and come to resemble closely the *evolutionarily stable strategies* that have so revolutionised our understanding of biological evolution.[12] The result of this sense of direction is a set of different goals or *futures* that have a sort of final cause quality in that they are projected 'out there' by the various desires that are socially generated in the 'here-and-now'. So this combination of socially-induced strategy and historical contingency is the source of the *goal-setting*, and the various ideas of nature provide the justificatory bases for the pursuit of those goals. In this way ideas of nature become *political* in the deepest possible sense of the word. This is because (thanks to the link that cultural bias provides between the realm of ideas and realm of actions) a person who acts in accordance with a particular idea of nature will be acting to strengthen the particular organisational form that receives its ultimate justification from that idea of nature.

Obviously, for policy analysis, this deep political aspect – the setting of goals in terms of preferred patterns of social arrangements, the *a priori* moral justifications for those goals, and the strategic bases for their pursuit – will be the main focus of interest within this cultural theory. Since the biases are *cultural*, and since their significance is *political*, it seems reasonable when handling them in this particular aspect to speak of them as *political cultures*. And finally, when those who variously hold to these political cultures act in accordance with them, they generate within the concrete reality – within the process of social life – the distinctive part-regimes that are the basic building blocks from which *political regimes* are constituted. In this way the concept of political cultures provides the essential, and currently missing, link between ideas and actions – between

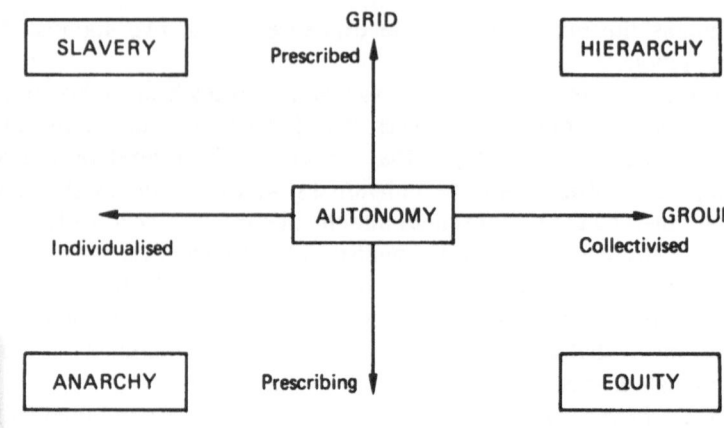

Figure 4.3   The five part-regimes

moral philosophy and political science (Figure 4.3).[13] We can now define political rationality. An act is rational if it supports one's political culture. And, conversely, any act that supports one's political culture is rational. Now we are in business.

Political cultures and the formation of policy

Of these five political cultures only three are likely to be active in any policy debate; the ineffectuals cannot gain access, and the hermits deliberately steer clear of all that sort of involvement. So policy debates are biased in the sense that two rationalities – the *rationality of fatalism* (slavery) and the *rationality of immediacy* (autonomy) – though present in the populace, are not represented in the debates. At their widest the debates will encompass just three rationalities – *market rationality* (anarchy), *bureaucratic rationality* (hierarchy), and the *rationality of truculence* (equity). Each of these, drawing on its appropriate strategy and idea of nature, projects its desired future 'out there' and then fleshes out into a living scenario the trajectory by which it must be reached. Like myths, scenarios work themselves out in men and, like hope, they spring eternal in the human breast.

Such scenarios, of course, are historically contingent (which is why they have to go on and on working themselves out and why they spring eternal) but at present they are nicely contrasted within the energy debate as the *business-as-usual* scenario (anarchy and market rationality), the *middle-of-the-road* (*technical fix*) scenario (hierarchy

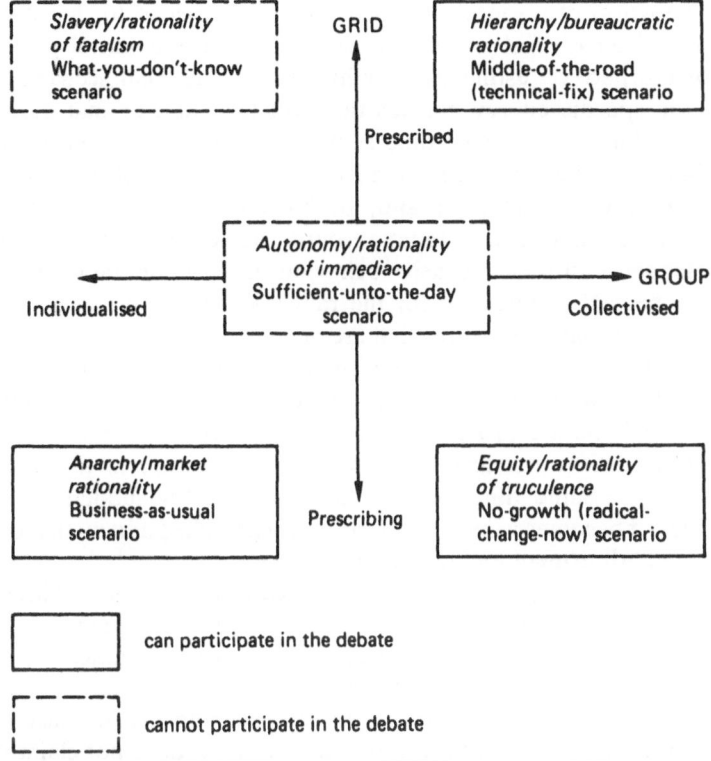

Figure 4.4   The political rationalities and their scenarios

and bureaucratic rationality) and the *no-growth* (*radical-change-now*) scenario (equity and the rationality of truculence) (Figure 4.4.).[14]

Justifying the scenarios

Our ecologist and our physicist hold their different ideas of nature, not because of differences between ecology and physics as systems of knowledge, but because of the different ways in which each of them is caught up in the social organisation of those systems of knowledge. The ecologist is a charismatic figure in a sect-like environmentalist group; the physicist is a 'Big Man' – in his youth, the Manhattan Project, more recently, the forceful leadership of a vast national laboratory. It is to these very different social contexts, and not to differences between ecology and physics, that their ideas of nature should be traced.

The world of resource abundance that is provided for him by his cornucopian idea of nature furnishes our entrepreneurial physicist with the perfect justification for his business-as-usual scenario and for the specific energy policies that will lead him to that glorious future. And, at the same time as it is justifying these policies, it is highlighting the nonsenses entailed in those rival policies that are striving towards different and, to his mind, less rosy futures.[15] In contrast, a world of rapid resource depletion – the inevitable consequence of such expansive behaviour in an accountable nature – is the perfect justification for the radical-change-now scenario and for the Draconian measures that are needed if it is to be reached . . . before it is too late. But isomorphic nature clashes with both of these. Discounting both resource abundance and resource depletion, isomorphic nature provides a world of resource scarcity. Resource scarcity justifies extensive government intervention in the market but, at the same time, rejects the argument for sudden change. Rather, it becomes a question of carefully controlled and meticulously planned adjustment and transition.

So these notions of resources are the carrots: the 'natural' induce-ments to act in certain ways and to advocate certain policies. And they are accompanied by the sticks: the 'natural' penalties that will be incurred if the carrots are disregarded; these are *risks*.

The cultural theory of risk begins by rejecting the literalist view that risks are objective but, at the same time, it does not claim that risks are all in the mind. All it says is that risks are selected and that there is a social basis to the resulting pattern of selection biases.[16] Risks are selected (and rejected) in such a way as to maintain a stable relationship between social context (organisation) and cultural bias (idea of nature). I can use the work of a political scientist, David W. Orr, to explain the different energy-related risks that gain salience in each political culture and the way they help to advance the hidden political agenda – the ushering in of a desired style of governance.

Risk for

Orr,[17] in trying to make some sense out of the energy debate in the United States, has identified three distinct *perspectives* each of which is appropriate to a set of primary actors and with each of which goes a preferred style of governance and a distinct set of salient risks. Each perspective, moreover, gains its particular orientation from the distinctive way in which the problem is defined. It is here, in the credible ways of defining the problem, that the different ideas of nature come into play, but before investigating that I should point out that Orr's scheme is impressively redundant in that each perspective's distinctness is defined over and over again by a whole list of different criteria – it is a *polythetic* classification.[18] For instance, he goes on to separate out the different energy goals that each perspective is striving towards, he lists the qualitative value changes that will be entailed, and he ends up with the different 'ultimate energy sources'.

In what Orr calls the '*Supply Perspective*' the problem is *inadequate energy supply*, the primary actors are the *energy corporations*, the preferred style of governance is *laissez-faire* – a minimum of government involvement – and the salient risks are those associated with *economic disruption*. In the '*Conservation Perspective*' the problem is *energy waste*, the primary actor is *government*, the preferred style of governance is *leviathan* – a major role for government – and the salient risks are those associated with *balance of payments*, *overseas dependence*, and *energy wars*. In the '*Energetics Perspective*' the problem is *social and cultural*, the primary actors are *the public* (I would prefer to say the *public interest groups*), the preferred style of governance is *Jeffersonian* – one in which a participatory citizenry blows the whistle on government – and the salient risks are *technological accidents*, *resource exhaustion*, and *climate change* (Figure 4.5).

The polythetic quality of this classification laces each perspective together into a whole package, as it were, and in so doing emphasises the unity of each and their clear separation from one another. I put the *ultimate energy sources* together in a separate box in order to emphasise that each package is assembled in such a way as to lead inevitably to the desired future, while each is so separated from the others as to constitute a *chreod* – a necessary path – that, once committed, cannot be changed. This does not mean to say that policy cannot hop this way and that between these paths but only that the three paths that between them define the policy space will always remain clearly separate.

One consequence of all this is that risk is never just risk but always 'risk for' (in the same way that history is always 'history for').[19] (The 'risks for' are the sticks – the sanctions – that are being used to drive the society towards the desired energy future and, more importantly, towards the desired pattern of social relations that is perceived as accompanying that future. Risks, in other words, are selected in order to provide rationalisations (in terms of the different rationalities

PERSPECTIVE / CRITERION	SUPPLY	CONSERVATION	ENERGETICS
THE PROBLEM	Inadequate supply	Energy waste	Cultural and social
PRIMARY ACTORS	Energy corporations	Government agencies	The public (public interest groups)
ENERGY GOALS	Inexhaustible cheap energy	Near term: efficiency Long-term: inexhaustible (but not cheap) energy	Decentralised solar-based society
PREFERRED STYLE OF GOVERNANCE	Laissez-faire	Leviathan	Jeffersonian
VALUE SYSTEM CHANGES REQUIRED	No change	Small (and gradual) change	Large (and sudden) change
SALIENT RISKS	Economic disruption	Balance of payments Overseas dependence Energy wars	Technological accidents Resource exhaustion Climate change
ULTIMATE ENERGY SOURCE	Breeder/fusion	Conservation leading to breeder/fusion	Decentralised solar, wind and bio-mass

Figure 4.5 Orr's framework

that inform each cultural bias) for preferred patterns of social relations. That, given the inevitability of 'risk for', is the cultural definition of risk.

Beyond self-interest

Most policy analysis approaches policy debates in terms of 'the decision-making process'. Such an approach begins (like Orr) by identifying 'the interested parties': the groups and individuals who, in pressing their different advocacies, give rise to the debate. Such an approach has to assume:

(a) that those who are not party to the debate are not interested;
(b) that the reason for the interest of the interested parties is self-evident – it is essentially self-interest;
(c) that what they are talking *about* in the debate is what they are interested *in*.

The cultural approach queries these assumptions rather in the way that 'the New Journalism'[20] queries the assumption that reportage (mere reportage, some die-hard positivists would say) is just some self-evident data-base from which literary creation then takes off. In querying these assumptions it has us ask some unfamiliar and intriguing questions:

(a) What of those who are interested but cannot gain entry to the debate, and what of those whose interest is best served by steering well clear of the debate?
(b) How do people who act in their own best interest come to know where that interest lies; that is, how are the goals they seek set?
(c) What about the hidden agenda; if all those parties are really arguing about something else – about what kind of society we should live in – should we not try to read the debate in those terms and regard its visible agenda as little more than a convenient medium for the expression of these social concerns?

These are the questions that cultural theory tries to answer.

Orr's framework is, of course, historically contingent: it is specific to a particular society (that of the United States) and to a particular period (the late seventies) in the history of that society. It certainly makes a lot of sense of the space–time context to which it is anchored and, in addition, it provides a tantalisingly suggestive orientation for understanding other debates in other places and at other times. The problem, therefore, is somehow or other to cut the adhesions that tie this framework to its unique historical context so that we can move towards an understanding of the eternal bases of which it is but one specific manifestation. How, in other words, do we move from phenomena to their possibility?

Orr's scheme is essentially an explanation in terms of goal-seeking; the goals being set by the evident self-interest of his primary actors. In order safely to cut it free from its anchorage in space and time we need to underpin it with an explanation at the much deeper level of goal-setting. We need to ask how it is that the primary actors can

SOCIAL BEING / CRITERION	ENTREPRENEUR	HIERARCHIST	SECTIST
ORGANISATION (CONCEPTUAL SCHEME)	Ego-focussed network	Hierarchically-nested group	Bounded egalitarian group
CULTURAL BIAS	Pragmatic materialism	Ritualism and sacrifice	Millenarianism Fundamentalism
SOCIALLY INDUCED PERSONAL STRATEGY	Individualist manipulative	Collectivist manipulative	Collectivist survival
IDEA OF NATURE	Skill-controlled cornucopia	Isomorphic	Accountable
CARROT JUSTIFIED BY IDEA OF NATURE	Resource abundance (culturally bestowed)	Resource scarcity (culturally bestowed within natural frame)	Resource depletion (naturally bestowed)
STICK JUSTIFIED BY IDEA OF NATURE	Economic risks (market)	Control risks (bureaucracy)	Involuntary and irreversible risks (voluntarism)
PART-REGIME	Anarchy	Hierarchy	Equity
SCENARIO THAT STICKS & CARROTS ARE STEERING TOWARDS	Business-as-usual	Middle-of-the road	Radical-change-now

Figure 4.6 The cultural underpinning for Orr's framework

come to know where the self-interest that they act in lies. But we have already answered this question – the explanation of goal-setting is to be found in the mutually reinforcing relationships between organisational types and ideas of nature. All we need to do is to slide this eternal cultural framework beneath Orr's historically contingent scheme (Figure 4.6). Again, this classification is polythetic and, again, I have put the scenarios in a separate box in order to emphasise the way in which each cultural/organisational package is put together in such a way as to lead inevitably to the desired future. Nor is this a complete framework: it is possible to go on adding more and more separation criteria, thereby adding to the strength of the separation between the three packages. The three participating rationalities, for instance, can be added and so too can Lakatos's anomaly-handling styles: *monster-accommodating* fits the expedient opportunism of the entrepreneur; *monster-adjusting* nicely matches the sorts of rearrangements that leave the hierarchist's essential frame unaltered; and *monster-barring* perfectly expresses the 'foreign body expulsion' that serves to maintain the sect's pure equality.[21]

The cultural hypothesis's polythetic classification, you could say, is like a tool-kit. You can go on and on adding to it and, though this may be a satisfying activity in itself, the main thing is to develop some sort of 'feel' that will enable you to select from it the most appropriate tools for each particular job of policy analysis.

IV CONCLUSION

The philosopher, Quine, once remarked that a linguistic or a psychological theory of the *a priori* – of our gut convictions about the nature of the world – would be 'a major philosophical achievement'. Interestingly, it did not occur to him that there might be a sociological explanation of the *a priori* but that, in fact, is precisely what the cultural theory provides. Even more interesting, perhaps, is the fact that this theory has originated, not in those prestigious and rarefied areas of intellectual enquiry where anthropology and philosophy traditionally meet, but in that grubby and pressing arena where the policy issues of the day are hammered out in often distressingly messy and unedifying debate. The cultural theory is an exercise in street-smart philosophy; its aim is to get to grips with, and change, the way we do things. In giving us certainties that are contradictory but not chaotic it stops us demanding to know which one is right and directs our attention towards the more realistic task of finding out how best to live with them all.[22]

Notes

1. I say 'our' ecologist and 'our' physicist because I do not wish to imply that each is typical of his profession. It may well be that more physicists think like our physicist than do ecologists (and vice versa) but, as I will explain later, the reasons for this are to be sought in the way their respective disciplines are organised and not in the disciplines themselves.
2. Human, not because animals and plants are not rational, but because they do not have (much) culture. Ideas of nature do not intervene between them and their environments.
3. Take, for instance, one of the well-known 'gambler's fallacies':

   > If some people believe that after a long run of heads the probability of tails on the next toss will be greater than 50 per cent, then one possibility is that they should be interpreted as believing thereby in a spirit of distributive justice that regulates the whole cosmos with a policy that ensures every-increasing probabilities of a trend-reversing intervention whenever identical outcomes begin to succeed one another within an otherwise chance set-up . . . a gambler's metaphysical belief may be at fault but not the rationality of his reasoning from it. (Jonathan L. Cohen, 1981, 'Can human irrationality be experimentally demonstrated?', *Behavioural and Brain Sciences*, 4, pp. 317–70.)

4. The argument for this sort of mechanism, and its plural consequence is set out in my *Rubbish Theory* (Oxford University Press, 1979) particularly in Chapter 7.
5. For the full theory see M. Thompson and P. Tayler, 'The Surprise Game: An Exploration of Constrained Relativism', The Aston-Warwick Scale Papers. No. 04. (1986). Available from Institute of Management Research and Development, University of Warwick, Coventry CV47AL.
6. M. Tribus, *Rational descriptions, decisions and designs* (Oxford: Pergamon Press, 1969) p. 2.
7. Kenneth Boulding (1983) 'National Defence Through Stable Peace' (lectures presented at IIASA, Laxenburg, Austria, June/July, 1981).
8. Mary Douglas (1978) 'Cultural Bias', *Occasional Papers of the Royal Anthropological Institute*, No. 34, London.
9. W. R. Ashby, *Introduction to Cybernetics* (New York: Wiley, 1956).
10. For a description of this context, see my 'The Problem of the Centre' in Mary Douglas (ed.), *Essays in the Sociology of Perception* (London and New York: Routledge & Kegan Paul, 1982).
11. For a rigorous treatment of this argument, see W. Brian Arthur, 'On Competing Technologies and Historical Small Events: The dynamics of

choice under increasing returns', Working Paper, IIASA, Laxenburg, Austria, 1983.

12. J. Maynard Smith (1980) 'Evolutionary Game Theory', in Claudio Barigozzi (ed.), *Vito Voltera Symposium on Mathematical Models in Biology. Lecture Notes in Biomathematics No. 39* (Berlin, Heidelberg, New York: Springer-Verlag, pp. 73–81).

13. Here I can do little more than mention the idea of part-regimes. For an initial development of this idea, and of the essential pluralism that it entails, see Aaron Wildavsky's *The Nursing Father: Moses as a Political Leader* (University of Alabama Press, 1984).

14. The names of these scenarios are derived from Peter Chapman (1975) *Fuels Paradise* (Harmondsworth: Penguin) who, in turn, derived them from *Exploring energy choices*, a preliminary report published by the Energy Policy Project, the Ford Foundation.

15. For a more detailed account of this see my 'Among the Energy Tribes: The Anthropology of the Current Policy Debate', IIASA, Working Paper, WP–82–59.

16. See Douglas, Mary and Wildavsky, Aaron (1982), *Risk and Culture* (Berkeley: University of California Press).

17. Orr, David W. (1977) 'US Energy Policy and the Political Economy of Participation', *Journal of Politics*, vol. 41, pp. 1027–56.

18. Redundancy does not mean that all but one of these criteria are unnecessary. That would be true only if the environment in which each perspective was being maintained was completely calm, and this is most certainly not the case here. Each perspective's environment contains the other rival perspectives and, in consequence, is turbulent in the extreme. In such an environment, redundancy is essential to viability.

19. Lévi-Strauss, Claude (1966), *The Savage Mind* (London: Wiedenfeld & Nicolson) p. 257.

20. Literary people were oblivious to this side of the New Journalism, because it is one of the unconscious assumptions of modern criticism that the raw material is simply 'there'. It is the 'given'. The idea is: given such-and-such a body of material, what has the artist done with it? The crucial part that reporting plays in all story-telling, whether in novels, films, or non-fiction, is something that is not so much ignored as simply not comprehended. (Tom Wolfe, 1973, 'The Feature Game' in Tom Wolfe and B. W. Johnson (eds), *The New Journalism*, Picador, 1975 edn., London, p. 27.)

21. Lakatos, Imre. 1976. *Proofs and Refutations: The Logic of Mathematical Discovery* (Cambridge: Cambridge University Press).

22. For an example of this see my 'Among The Energy Tribes: A Cultural Framework for the Analysis and Design of Energy Policy', *Policy Sciences* vol. 17, no. 3, Nov. 1984, pp. 321–39.

5 Technology as Cultural Process

Brian Wynne

I INTRODUCTION

Amongst the various approaches to analysing people's attitudes to risks, technologies and their managing institutions, which have developed such a rich texture in recent years, one can see two fundamentally different metaphysics from one or the other of which nearly all approaches originate. By metaphysics I mean a closed loop – a cosmology – of taken-for-granted views of human nature and social interaction, of public life, rationality, values, and of ways of observing them which select those aspects which confirm our founding assumptions and faiths.

Thus many modern analysts take it for granted that individuals are discrete beings, morally insulated from their social context and constituting integrated values which are complete, and more or less stable. People's values are assumed to be scientifically observable, solid, clear (or, at least, capable of being 'clarified') and factorisable. This usually also entails the view that people interact according to clear-cut self-interests which they are busy optimising. Social life is just a calculus of rational, individual, self-interested, material optimality. The methods of observation of these aspects of human nature and social behaviour reflect and confirm their basic metaphysics.

An alternative approach – which I confess to finding more congenial and plausible, and which I shall use in this paper – takes the opposites of all the previous aspects. Thus people are assumed to be *intrinsically* social, down to their very identity and 'values'. These are made up *by* social interaction (which does *not* mean they are plastic to present social interactions, to the neglect of history, socialisation, basic elements of cultural context, and so on). Thus, in this approach, human nature is always essentially incomplete, and values are intrinsically vague, unfinished and available, to some significant extent, to social negotiation and definition (which is sometimes called 'clarification'). Seen thus, ambiguity in values can be seen as itself a rational defence against being inflexible and pinned down when social and other uncertainties would rationally require flexibility. The methodological counterpart is that 'scientific' methods, for all their power and precision, capture only one dimension of the way values and attitudes form. Furthermore, they may be no less laden with essentially unproven and metaphysical commitments than other approaches which may, like the anthropologist's, draw upon a richly varied repertoire of sources of data. The apparently ill-disciplined, often anecdotal character of such observations and data may be equally worth being taken seriously as the 'scientific' methods more fashionable in consultancies and so on. Borrowing the words of a famous beer advertisement, this approach may reach the parts that other brews never dreamt existed.

In this paper I will attempt to develop the recent argument that rigorous pursuit of the perception of risks leads us towards a political and cultural view of technology as a social institution and indeed as social *process*. I will then relate this to a sociological, even psychoanalytical, interpretation of attitudes by suggesting that we view technology as a cultural process. I argue that only with this analytical approach to technology can we take seriously the question of public perceptions of risks in technology-dominated societies.

II TECHNOLOGY, NATURE AND CULTURE

Just as there are political implications in the way nature is defined, so too in the way 'technology' is defined. Although there is an apparently irresistible urge to use terms like 'nature', 'culture' and 'technology' as if they were unitary entities (and perhaps that is always the fate of potent social symbols), public policy reflection is better served by examining the origins and implications of received definitions and their 'interfaces'.

There has been a long tradition of research on the social negotiation of nature and its complex relationship with culture.[1] What Thompson refers to as an eternal circle, of the cultural construction of nature and the natural destruction of culture,[2] leads to the apparent conundrum of the cultural destruction (via nature) of culture. The conundrum only appears, however, if we give the floor to the received approach to 'Culture', which is to see it as a homogeneous, monolithic whole – Western culture, Islamic culture, traditional culture, and so on. Adopting a more modest notion of culture we can attend to the contending differentiations within (and elements of cross-'cultural' identity between) such abstract monoliths, and link these to real beings, institutions and issues rather than moral ideals. We can see different social groups, their characteristic customs, belief systems, social interactions, as more or less discrete local cultures, maintaining their own identity and existence in relation to others, within the larger melting-pot. Cultures are at the same time destroyed as active social constructs and yet immortalised by being 'naturalised' by their proponents.

The distinctive essence of 'culture' as a framework of analysis is the integrated wholeness of cognitive and material social dimensions of existence. Nature is worked on and manipulated through ideas of nature, society and technology which correspond with basic patterns of social relationships in that 'culture'. This is not at all an approach antagonistic to conventional notions of social structure, institutions, power and economic relations: rather it may enlarge our vision of how such material social realities are maintained or changed.[3]

In this paper I want to explore the notion of technology as cultural process in this sense, of embodying a differentiated set of cultures, each of which may be essential to the technology, but between which relations of power, communication and co-ordination may be problematic. It could well be asked why one should use the term 'cultural' rather than 'social', and I have indeed tried elsewhere to suggest the practical importance of seeing technology as social organisation.[4] However, without wishing to deny the importance of organisational, economic and physical elements of technology, I am trying here to emphasise the associated attitudes, images and belief systems which legitimate the social relations of technology. The ultimate goal is to shed a little light on the complex, brittle relationship between alienated 'acceptance' and active attempted involvement (often via protest of some form) in the social direction of technology. This relationship is a key node in the dynamics of social and technological change and perhaps in the historic project to re-embed technology in more democratic forms of control, but it is at the same time obscure and highly unstable. It is worth studying by methods less regimented than orthodox attitude surveys can offer.[5]

Although the relationships between technology and culture have long been a topic of inquiry,[6] the notion of technology *as* (differentiated) culture has been of far less concern. Suggestive but neglected work by Edge a decade ago on some cultural implications (for example, the 'dehumanisation' question) of technological metaphor acts as a springboard for my explorations.[7] I will attempt to relate some of his insights to recent work in cultural anthropology as developed to address some modern policy issues concerning technology.[8]

First, however, I will outline a schema for treating culture more realistically, as a differentiated context of competing social–cognitive–metaphysical styles.

III TOWARDS POLITICAL CULTURES

Several cultural anthropologists associated with Mary Douglas have developed an essentially two-dimensional framework of socio-cultural attributes by which to define basic structural differences and comparisons between cultures. This 'grid–group' comparative classificatory system may be applicable at various levels of aggregation from 'national' cultures to individuals within sub-cultures. It has been well articulated elsewhere,[9] so that only a brief outline is needed here. My aim is only to use the framework as a way of seeing in context the relationship between passivism and active protest. This may, in passing, go a little way to adding some needed social dynamics to the framework itself, but that is not my main objective here.

The approach starts from the not unusual premise that ideas of nature held by groups and individuals correspond with basic moral principles crucial to such a group's self-maintenance. Egalitarian groups tend to 'naturalise' and thus maintain moral equality by seeing biological equality in nature. Hierarchical groups would tend to reflect their social hierarchy in perceptions of hierarchical processes in nature. This is standard fare. Cultural filters shape the perception of nature in systematic ways, blocking inconsistent data and highlighting confirmatory data. These filters are not merely encrusted habits learned by rote and mindlessly enacted from one generation to the next; they are the product of active scheming to maintain a given cultural style or bias in contention with competitors. The theory of the cultural anthropology school associated with Mary Douglas is that from all social contexts there are only a few fundamental types which such cultural biases can take. These can be mapped on orthogonal axes of 'grid' and 'group'.

High-grid social relations involve a high level of external social prescription of the 'individual's' role. (More correctly, the role of the social unit, which may be a group or individual.) There is little or no autonomy, and the actor's experience is as a manipulated periphery to someone else's centre. Conversely low- (or negative-) grid relation involves high autonomy and anomie, and strong prescribing towards others.

High-group relations involve strong incorporation into sharply-bounded groups. This social demarcation between inside and outside is the key property. 'Negative group' would mean active rejection of group boundaries.

It is the orthogonal combinations of these properties which provide empirically recognisable social groups, individuals and organisations. Thompson has described them as follows:

> The group and grid axes have both positive and negative directions. Since group and grid can only be measured on ordinal scales, there are only five distinctions to be made within this social context space – one at the origin and one in each of the four quadrants. In each of these distinct social contexts we find a distinct social being: at the centre, *the hermit*, free from coercive involvement in both group-formation and personal network-building; at the bottom left, *the entrepreneur*, spurning group involvement and central to a large personal network; at the top left, *the ineffectual*, excluded from social groups and peripheral to the personal networks of others; at top right, *the hierarchist*, strongly grouped and willingly subject to all the prescriptions that serve to maintain the ranked separation of his group from all the others within the group hierarchy; and at bottom right, *the sectist*, strongly grouped but rejecting hierarchy and all the prescriptions that are its inevitable accompaniment.

> I trace these five stabilisable conjunctions of social context and cultural bias back to three distinctive kinds of organisation: *the ego-focussed network*, *the hierarchically-nested group*, and *the bounded egalitarian group*. I further argue that this typology of organisations is exhaustive – that these are the only kinds of organisation that are socially viable.

This scheme is represented in Figure 5.1, where illustrative labels are given for the five basic social types, their typical cultural biases, moral justifications and ideas of nature.

If we apply this scheme to ideas of technology we can see corresponding patterns. A high grid view would emphasise highly structured forms, and a high group would emphasise strongly bounded

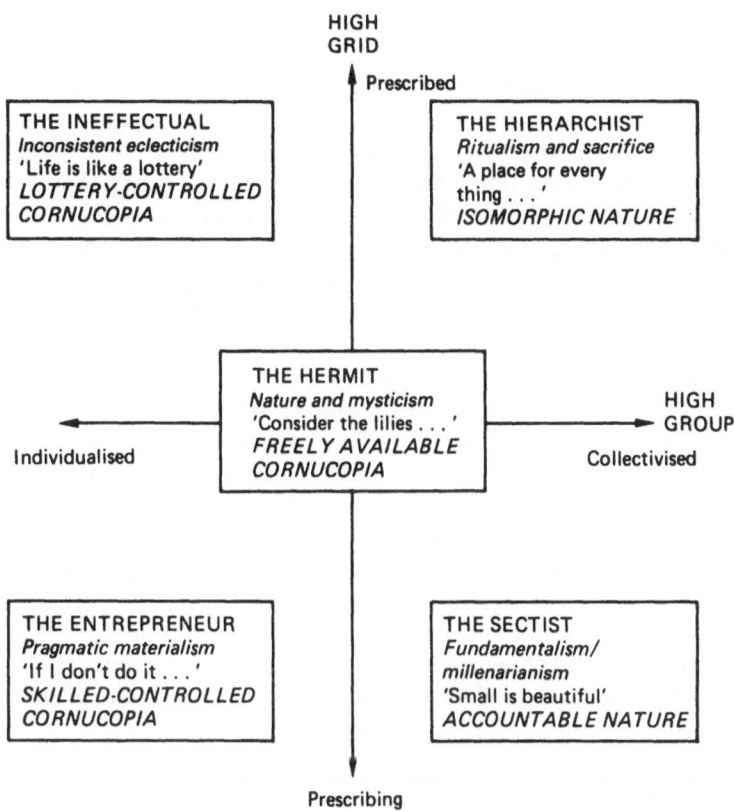

Figure 5.1 The grid-group typology of cultural styles (Thompson, 1983): SOCIAL TYPES, *cultural biases*, 'justifications' and *IDEAS OF NATURE*

areas of technical control or consequences, that is, strong boundaries of responsibility. Thus a combination of high grid and high group would yield a sense of well-ordered technical action with, in principle, clear-cut boundaries of consequences. If these are not actually clear-cut then better forecasting and assessment can achieve this. Hence there is a sustained concentration bordering on the obsessional, with refined techniques for technological forecasting, risk management and technology assessment. High grid–low group, on the other hand, would yield an analogous sense of determinism in the direction of technology; but this is an inaccessible determinism, out of reach of recognisable, organised human perception and management. Technology is well gridded, but without any recognisable moral community (group) because there is no strong group experience to frame that sense.

Low grid–low group, on the other hand, would entail a similar sense of unpredictability to the low group top left of Figure 5.1, but this time an unpredictability that was accessible, thus open to exploitation – sometimes this would pay off, at other times not. Anticipation would be of limited value; it would be more a case of 'ride the tiger' than manage in the conventional sense of cross-impact matrices, nth order consequence probabilities, and the rest.

Finally, the high group–low grid style would emphasise strong boundaries of responsibility, discontinuous consequence profiles (apo-calyptic tendencies), and a low sense of external determination; that is, a high moral responsibility to direct technology, but in more collectivist ways. Hence there would be an emphasis upon normative management, but more via collectivist styles of political organisation ('appropriate technology') than by conventional hierarchical forms of management. From the point of this bias, technologies would tend to be evaluated according to perceived intrinsic moral qualities.

One can see how these ideas of technology tend to correspond with ideas of nature. Indeed within each cultural style the ideas of nature and technology interpenetrate and reinforce one another. There is no clear boundary between nature and technology: indeed our publicly certified knowledge of nature, namely science, is now-adays certified only via technology, that is, as knowledge leading to greater technical control, and nothing else. Truth and manipulation have become culturally confused.

The existence and character of the *hermit* type is subject to some conflict: and all the basic types are seen as mixed and nested in social reality. The grid–group classification can be applied at various levels of aggregation, from the individual, to specific groups, to whole occupational types, to national political cultures or whole societies. Although this has occasionally been treated as a sign of inconsistency in the schema, it is more relevant to view it is classificatory rather than of itself explanatory. It is a necessary preliminary to explanation. The 'problem of levels' then becomes less serious, and indeed may be a positive source of development of the approach towards the more complex question of social change via the interactions of such basic types. For example, single organisations may contain a rich blend of entrepreneurs, hierarchs, sectists and ineffectuals. Within an overall hierarchical, formal, organisation, sectist groups may emerge and operate at a given level, say in response to moves to reorganise or discontinue their work. Entrepreneurs (formal and informal) may also operate at different levels, in constant tension yet overall unity with the organisation as a whole.

Whatever the difficulties of consistent applicability and empirical referents of this, itself rather fundamentalist, schema, it does appear to resonate with broader experience and research on organisations and (with more difficulty) political cultures.[10] A further criticism, however, is that the basic metaphysics of this theory are a version of 'naïve pluralism'; that although the cultural emphasis usefully reintegrates cognitive dimensions of social behaviour, there is no talk of power, even though the schema claims to encompass political affairs.

A valuable linkage has been provided, however, by Thompson's suggestion[11] that the diagonal between *hierarchist* and *entrepreneur* (bureaucrat and innovator) can be regarded as a joint axis of power (and complacency) as conventionally treated in social science. In 3-D space, with power as a third dimension, this axis could be regarded as a ridge connecting the entrepreneurial quadrant with the hierarchist. Although tensions exist along it, there are many coalitions and elite formations in society which constitute this ridge system. The *ineffectuals* can be regarded as a pretty flat landscape, and the *sectists*, for our purposes radical grass roots labour union sections, environmental or other activist campaigning groups, a slightly less lowly and, as we shall see, more turbulent landscape. This diagonal axis might be called the axis of instability and powerlessness.

Although this rough schema gives us the opportunity of testing ideas about changing social patterns through the whole system, I am interested here in exploring only one part, namely what makes people and groups move from being passive, alienated and disoriented 'ineffectuals', to become active, even zealous, interveners in the process of technological decision and development. How does this apparently unpredictable, sudden process come about? This has become a question of great practical importance, whether to government agencies wishing to anticipate and contain such movements within their planning horizons, or to activists wondering why they are not being joined by mass uprisings in their cause. Dissatisfied by the simplistic (though no doubt partly true) NIMBY explanation,[12] I have tried to dig deeper into the labyrinth of psychic tunnels by which ineffectuals, rather than try to scale the ridge separating them from activism, instead burrow through like moles to the other side.

IV THE AXIS OF INSTABILITY

Although the ineffectuals category is probably the most complex of all those advanced by the grid–group schema, this complexity renders it perhaps the most interesting. By definition, many of the attitudes and beliefs of this category are inarticulate, partial and latent. This is, after all, the central arena of the perennial 'false consciousness' question.[13] All the other groups use this passive, if differentiated, majority in their own schemes, involving different versions of 'the public interest' and different theories of *why* the majority is so silent, corresponding with their own cultural bias. This sector could perhaps be regarded as a heterogeneous, aqueous solution, invisibly super-saturated in parts, where local seeding gives sudden crystallisation and an entirely new constellation of phases and interactions. These new phases are our analogy for activist groups with egalitarian sectist properties.

From beaming its ideology one way towards the passive, alienated majority, the axis of power now suddenly has to face the opposite direction too. To maintain power and authority towards this sector with its different rationality may require very different ideological contents, perhaps even ones contradictory to those effective for keeping the ineffectuals quiescent. This is suggested conceptually in Figure 5.1, and is borne out in empirical experience.

For example, when controversial policy decisions about complex technological developments are made by institutions such as public hearings, legal processes, and so on, they are usually described in the public language as expert discovery problems. This description only inflames the (often well-informed) relevant activist groups, because their disaffection is strongly rooted in antagonism towards expertise and technocracy. They demand more explicit recognition of moral and political choices – a (low grid) language of prescription rather than objective structure. What is good legitimating language for keeping the quiescent majority quiescent is exactly the opposite for these 'sectist' activists. I have elsewhere described in detail this tension in the case of the 1977 Windscale Inquiry and its framing legal rationality.[14] Conversely, describing the issue in the language of inevitable expert uncertainty ('difficult, so you may lose') would have mitigated the impact on many activists, even if the specific decision had gone against them, because this language caters to their cultural style of moral prescription; but, by the same token, it would have invited some quiescents to join the fray and take issue.

Seeing this relationship as a fragile balance-in-tension of contradictory ideological tendencies and relationships offers us an analytical
framework within which the relatively sudden shifts which are
frequently seen in attitudes and levels of conflict – political surprise –
can be conceivable. Regular symbolic action[15] beamed in one direction
and apparently successful at keeping consensus-by-quiescence may
conceal from the view of the power elite the growth of activists as it
were popping up threateningly behind it. The cultural filters of the
elite may allow the activists to develop into significant features of the
political landscape, with solid connection (for example, via the skilful
use of the media) with the popular culture called 'ineffectuals', before
they begin to take them seriously.

Once they are taken seriously, however, some interesting dynamics
may emerge. Sectist groups are highly egalitarian, grass roots in style.
They are antagonistic towards leaders, spokespersons and experts,
which is why the leadership rituals of such groups are often more
agonising and bloody than those where at least the notion of a leader
is accepted.[16] Despite this, however, being taken seriously demands
that leaders and spokespersons be deputed. In regular necessary
interaction with the institutions of power such a role demands
increasingly expert, technical language of argument. Typically such
leaders move, in language, attitudes and style, towards the axis of
power and the elites they begin by rejecting. They find themselves
torn between, on the one hand, loyalty to their fundamentalist,
uncompromising grass roots with attendant bosom-like security but
politically 'outsider' status; and, on the other hand, incorporation
into the respectable margins of the policy elite, where status and
recognition are traded for the willingness to emasculate original
arguments into the narrow technical discourse controlled by the
establishment. They gravitate towards top-right, Figure 5.1, towards
co-option.

This kind of metastable state can exist for years, with activist
groups in a continual state of crisis and upheaval over their leadership,
proper strategies and styles of argument. If their leaders reduce
this tension by becoming too co-opted,* too drawn towards the
hierarchical sector, such groups may simply, and quite rapidly,
dissolve back into anonymity and the majority resume their member-

---

* This co-option process, its successes, ebbs and flows, depends also upon the structure
and flexibility of the establishment, which is at least partly relative to the specific
issue. Co-option may be more likely with confident establishments (such as the United
Kingdom) and less with insecure, thus intransigent, ones.

ship of the ineffectuals; the process of high group boundary-mainten-
ance and the fervent articulation of common identity and purpose
fall apart. It may be, of course, that co-option of leaders and the
emrgence of new leaders from the grass roots is an endless process,
maintaining the active if turbulent existence of vigorous 'sectist'
groups. If such groups do disappear from view, their members may
still make up a latent nucleus – to use the earlier metaphor, a super-
saturated area of solution – for later reactivation, perhaps on an
adjacent but not identical issue.

This kind of analysis corresponds strongly with the approach to
attitudes and behaviour which rejects the rational economic individual
calculator model, of values, goals and interests as clear, stable and
concrete.[17] It supports the view of people and their attitudes as
more tentative, experimental, incomplete and perhaps internally
inconsistent; humans as flexible managers of the conflicting, complex
grounds of their own being. They may be more ambivalent, 'unstable'
and open to suggestion of their goals and values by dominant
cultural stimuli than more individualistic, rationalistic approaches and
methods claim. It is culture, not individuals, which gives these values
what consistency and force they may have.

Since technology provides potent experiences and images which
shape meanings, perceptions and behaviour, it may be regarded as a
key substrate of culture. To the extent that modern technology
provides uniform mass experience, it may be a form of common
culture cross-cutting and underlying or destroying the differentiations
which the grid–group approach posits.[18] I shall advance this perspective: that such differentiations are under-acknowledged and are far
stronger than generally assumed; that the experiences, relationships
and their guiding rationalities which people invest in technology are
more varied, contradictory and important than received models of
'technology' can accommodate; and that their systematic analysis and
recognition is of practical importance to technology policy. It is worth
attempting to examine the expression of attitudes and self-images in
technological experiences, and then to explore the psychological and
sociological undercurrents of these. Technology can be regarded as
culture in the sense that it is a potent framework relating dimensions
of belief and meaning to social relations and processes; and it is
political culture in that the social relations of power embodied by the
technology are more or less successfully legitimated by the cognitive
structures which are naturalised in the culture, and which thus conceal

those underlying structures of power from critical examination and possible change.

V TECHNOLOGICAL ANIMISM

In his classic account of the social and psychic devastation caused by the 1972 Buffalo Creek dam failure in the Appalachian mountains,[19] Kai Erickson observes that the reaction of the economically and politically marginal people who were victims of that 'point-disaster' was profoundly conditioned by their internalisation of the state of 'chronic disaster' represented in their long-term neglect and alienation from employers and public authorities. The psychic withdrawal characteristic of extreme traumatic shock was already consolidated on the community scale in the alienation and self-dependence of the community, trusting none of the agencies on whom they nevertheless depended and whom they thus tolerated for economic survival. Erickson argues that what was most significant about the social aftermath of the disaster was not the personal trauma – 'psychic numbing' – which everyone experienced, but the collective trauma, the inability of the old social networks to re-establish themselves as the framework of personal psychic convalescence and development. The people felt betrayed by the coal company which neglected the dam whose burst caused the disaster, not because they had previously thought it a conscientious company, but because structurally, in their position, they had to trust it, despite realistic appreciation of its selfish motives, past neglects, and so on.

In Erickson's perspective[20] the powerless always tend to defend and rationalise, thus to *consolidate*, their own impotence and apathy because to do otherwise is to expose themselves to the greater human damage of *explicit* neglect and powerlessness. They withdraw, and justify and defend that withdrawal as being consistent with cosmic principles; it becomes their culture, integrating their beliefs about cause and effect in the experiences they encounter, with their established social relationships. Erickson saw the classic symptoms of trauma in the ordinary human reactions to 'the age we are entering', namely 'a sense of cultural disorientation, a feeling of powerlessness, a dulled apathy, and a generalised fear about the state of the universe'.[21] These correspond with the features of the 'ineffectuals' of the high grid–low group cultural category. They are the symptoms of social experiences and roles which are highly prescribed by others, yet where the structure of such prescriptions – of their own marginality and manipulation – is obscure. The 'effective causes' of their powerlessness are socially invisible. What Erickson also saw being enacted in social reality was the tentative, fragile nature of movements out of apathy and disorientation. What community developments there had been in that direction were swept away by the flood, which was analogous to the condensation into a single, extreme dramatic event, of years of non-affirmation (identity-stripping) by the outside world.

I have made this excursion into Erickson's interpretation of a man-made disaster in order not only to help uncover the complexities of attitudes and some continuities between historical events and historical processes: I also want to explore the way technology – here a dam central to the community's existence (it was part of the local coal-mine which employed most of the people) – is externalised in images which shape cultural attitudes that implicitly reflect back people's social relations as alien objects, beyond their control or responsibility to alter. In the Buffalo Creek case, the survivors seemed to have a clear sense of who was responsible for the technology's havoc, but an equal sense of hopelessness that anything might be done about it. *The effective cause of their disaster was at least seen as human agents*, even if these were believed to be beyond control. This, however, might be taken as an extreme example, of a concretely visible technology with clear lines of control and responsibility. Many other technologies typical of the modern age – nuclear power, genetic engineering, and perhaps most especially computers – lie at the opposite end. Their controlling human agents are invisible, diffuse and socially remote. It is impossible for ordinary people to identify the effective causes of their confusing and often troubling experience of these technologies, even if they do not produce dramatic interventions in their lives. Yet the importance of these experiences requires that people construct some working explanations so as to rationalise them one way or the other.

One example of the way such effective causes in technology have been mystified, and images cultivated, is given in Figure 5.2.[22] The technology – here a nuclear power station of the most 'advanced' sort (the Dounreay fast reactor) – is deified to the extent not only of hovering, disembodied, above the mere earth, but of having a halo to denote its moral purity and magical power. The caption invites people to awe-struck worship, dazzling them to any perception, let

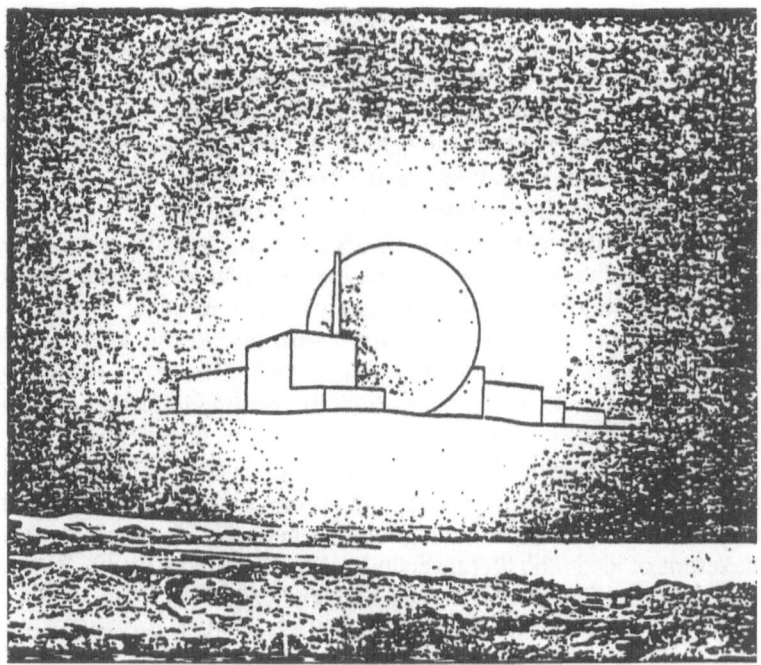

DOUNREAY — *to mankind will be the reward* . . .
Millions of people, with only a glimmering of what the Atomic Age
can mean to mankind, stand amazed at the fantastic prospect of heat,
light and power issuing from a source that cannot even be seen.
These people of the world, for whom the atom will be harnessed for
service, gather their news of this great new power, day by day, from
newspapers, radio and television.
These men and women can feel a sense of pride in the achievement
of harnessing this power for peace.

Figure 5.2
Source: *Financial Times*, London, October 1956

alone questioning of the agencies, interests, uncertainties and human
frailties behind the image. This is symbolic action *in extremis*.

An important consequence of this socially constructed invisibility
of effective causes in technology is indicated in a small part of the
caption of Figure 5.2. Part of the imagery of magic power is the fact
that the 'fantastic prospect' will emanate mysteriously, from a superior
force that cannot be seen, heard or felt. These properties of ionising

radiation, then used to intensify the positive power of the technology, are the very ones which are now regarded as intensifying exaggerated hostility and fear. In other words, legitimation was created by cultivating the idea of awesome, other-worldly power, beyond the bounds of ordinary nature and culture; but this disorienting relationship's corollary is a double-edged instability which can easily and suddenly flip over from *benign* to *malign* externality.

The point is that with effective causes and structures of responsibility so obscured, the only responses possible are *total* acceptance (tinged with an ambivalent potential for anxiety in the face of such supernormal power) or *total* rejection (tinged with fascination at the sheer technical mastery such technology may entail). There is no possibility for measured criticism and conditional, qualified responses – all possible currencies of discrimination have been historically obliterated, leaving behind inflexible absolutes. This is tantamount to primitive thought, where the symbol is collapsed into the word, and no creative tensions any longer exist between the metaphorical skeletons of ideas and literal versions of the metaphor. People behave as if the technology were literally an alien being from space.

Psychoanalysts have examined clinical cases involving similar condensed images of technology which have become central surrogates for explanation of more complex experiences and potential responsibilities which people cannot handle. These images, or *spectres*, are not only psychic simplifiers but also metaphors of social relationships; and they are built around technological images, perhaps increasingly so, given the increasingly central role of such experience in daily life. Daly defines a spectre as a kind of potent, artificially created but invisible behavioural force.[23]

A sense of the operation of such forces arises when men find they cannot account for emotionally significant events by ascribing them to the conventional sources of power and efficacy (e.g. human, natural, divine) which are believed to make things happen in the world. When such inexplicable events persist and are experienced by numbers of people, agencies are created to account for these events. These agencies are given names, made into realities, and adapted to as powerful things. . . .

The spectral view of technology arises from a sense of domination by mysterious forces or agencies which are, or were, linked to technological enterprises but which are now apprehended as being beyond the control of any particular man or collection of men. . . .

[People] behave as if the spirit of meeting specifications in many discreet, limited and finite human ventures had taken flight from the hands of responsible agents and become an independent reality – a reality which has come to overhang the modern world and to enter into the dynamic processes of personality – as a spectral object. (Daly, 1970).

There is, in other words, a ritual defence mechanism – a transference of responsibility for complex and inexplicable experiences which are too emotionally important to be ignored. Daly describes how several patients created such spectres of their own biological systems, investing them with powers to decide and cut a clean swathe through otherwise overpowering ambiguities. Thus they would obsessively refer to a simple measure such as their pulse rate as a guide to decision making – it was made into a source of 'objective decision rules' supposedly reflecting a greater, more powerful but impenetrable biological mechanism. Such agents may become absorbed into part of one's very identity; or more accurately perhaps, one's identity may be shaped by, then absorbed into the image; one becomes 'a cog in a machine' or, with Bettelheim's Joey, 'a mechanical boy', an electrical appliance who 'plugs himself in' and 'switches himself on' before he can speak, and who causes others to behave in parallel fashion in order to relate to him.[24]

It is a central point of Daly's analysis of these conditions that they are no longer, if they ever were, restricted to clinically psychotic individuals. They are now in his view mass neuroses, transmitted in normal processes of cultural dissemination. Given the kinds of symbolic action depicted in Figure 5.2, this is hardly surprising. Indeed the historical use of images of scientific, technical power as if from outside the realm of human interests and values has ironically cultivated an escalating search for objective decision rules from science, akin to a collective scale version of consulting pulse rates, such as the eternal effort to avoid the ambiguity of *negotiating* acceptability from situation to situation by instead creating objective scales of 'acceptable risk'.[25] The artificiality of these entities and the impossibility of their ever providing what they promise may never be widely apparent since they are embedded in a whole labyrinth of dense managerial political language and institutional barriers. But their constant usage in keeping the ineffectuals at bay is just what disaffects and activates the sectists even more.[26]

VI TECHNOLOGICAL ANIMISM AND SOCIAL ACTIVATION

In many cases the creation of such technological spectres may, ironically, be an essentially rational reaction to irrational situations. Most people are fragments of technological systems which entail many connected parts whose co-ordination is essential, but complex and chronically problematic. However, they never experience the whole system:[27] their experience is fragmentary and bounded by their local organisational and cultural context, within which they have to make out. Finding it impossible to penetrate the boundaries of their local experience and to understand the rationalities, interests and interactions of those whose doings structure that situation, they create shorthand images to 'explain' those external agencies and their frequent unpredictability and apparent malevolence.

A graphic example of this was given by McDermott, who described a spectre created by American GIs in Vietnam.[28] They were operating in the jungle, constantly sniped at or attacked by Vietcong guerillas who could never be identified and pinned down; they were regularly shelled and rocketed, but never sure it was not their own side; and received orders but never explanations from their superiors. Their experience was frightening, confusing, contradictory and utterly obscure as to its effective causes. They could not find an enemy and they could not identify their own side. Yet they received orders and were attacked in equally arbitrary fashion. As part of their rationalisation of this (very high-grid) predicament the GIs had condensed the potent, but diffuse and invisible effective causes into a single agent, a 'huge f...ing' gun which lived in a hollowed-out mountain, and which emerged at whim to unload death and destruction onto them. It was an agent beyond control, imbued with a kind of autonomous, malevolent intelligence. In one major sense it was no comfort at all, but in another sense it was, because at least it offered *explanation*. It was a kind of metaphor representing their social relationships with those elites (and here also enemies) who remotely and invisibly controlled their fate.

Langdon Winner has also discussed this process as technological animism.[29] He takes the story of Rudy in Vonnegut's *Player Piano*. Rudy was a mechanic whose job had been replaced by automation: his skills and experience had been reduced to an algorithm and entirely handed over to a computer. Deeply upset and mystified by this shattering of his very identity, Rudy enacts a scene in a cafe with a doctor friend, in which he goes into a frenzy over what he sees as the creepy, superhuman intelligence controlling the keyboard movements on a simple slot machine (Player) piano. Perhaps indicating that, as a more educated being, he sees through this conversion of concrete, if hidden, human goals and interests into extra-human, therefore untouchable intelligences, the doctor friend has to get up and walk out on this pathetic scene. In Winner's words,[30] this scene gives:

> a glimpse of the crucial statement and ultimate conclusion of the writings on technological animism. If one asks, Where did this strange life in the apparatus come from? What is its real origin? the answer is clear: it is human life transferred into artifice. Men export their own vital powers – the ability to move, to experience, to work and to think – into the devices of their making. They then experience this life as something alien and removed, something that comes back at them from another direction. In this way the experience of life becomes entirely vicarious. . . .
>
> Man now lives *in* and *through* technical creations. The peculiar properties we may notice in these creations are not the result of some spontaneous generation. What we see is human life separated from the directing, controlling, positive agency of human minds and souls. (Winner, 1978).

Winner's important insights here must, however, be qualified, or perhaps clarified, by one important point. Although men do 'export their own vital powers' into the technologies they have created, and reflect them back as aliens beyond control, this falsely implies a lack of any social stratification or cultural differentiation in this process. Elites are also immersed in their myths and fantasies about technological power, and non-elites do produce technological creations. But it is also important to see that those ineffectuals are circumscribed by mystifications created through domination by decision-making elites, a domination whose arbitrary human structure is increasingly socially complex, remote and thus 'invisible'. They therefore transfer responsibility from this frustratingly intangible and impenetrable human complex, onto extra-human spectres. This is transfer and condensation not so much of their own (anyway small) responsibility and power, but of the power of elites in the social structure around them. The myths and fantasies of the axis of complacency (see Figure 5.1) actively promote this mystification and concretisation of their own power, even if not always deliberately.

Not only does this cognitive process artificially consolidate the axis of power by placing it apparently beyond human access, but it encourages a lack of human tolerance for ambiguity, thus a structural brittleness in the system. When responsibility is so condensed into such technological spectres whose inner workings are inaccessible, experience has to be interpreted, and life conducted, either by total identification with or total repudiation of such spectres. Thus public 'debate' and interaction become rigid, and prone to sudden discontinuities: government itself may become less viable. As Crozier has put it, there is no authority without negotiation,[31] and since such fantasies and spectres pre-empt the possibility of negotiation by replacing and 'black-boxing' more discriminating perceptions of relationships and causes, they tend to destroy even the *possibility* of legitimate authority.

A good example of the absolute contradiction in different social perceptions of technology, and the linkages between these and power structures, arose during the 1977 Windscale Inquiry.[32] This was a public inquiry into a plan to build a new plant to reprocess spent nuclear fuel from the new generation of reactors, using oxide fuels. This would extract plutonium which could be used in fast breeder reactors or weapons, uranium which could be recycled in further thermal reactors, and radioactive wastes which would ultimately need some safe final disposal. The plant was part and parcel of a longer-term historical vision of nuclear development reaching out of colossal past commitments and into future ones. Its go-ahead naturally made all of those future envisaged commitments more likely, via institutional momentum and technical–economic logic.

The proponents and the High Court judge in charge defined the issue as the examination of the direct impact only of the reprocessing plant itself, and excluded any question of the implications of future fuel cycle developments which might be entailed by it. These, he argued, would be subject to future separate decisions, and any attempt to cover more comprehensive nuclear futures was 'emotive nonsense'. Yet many objectors took it for granted that the reprocessing plant, being only a part of a historical process, had to be examined as such. Fast reactors, plutonium trading, waste disposal, further reprocessing plants, and so on, all had to be considered.

This issue was only 'resolved' by the fiat of the judge. He found it impossible to negotiate with this alternative definition of the problem, perhaps because it was rooted in *objectively* different social experience, which he defined as 'merely' emotive. To the decision-making elite it was logical to say that future plants could be separated as decision issues, because they could identify with the whole process in which those future decisions, as well as the present ones, would be made. They could conceive of decision choice and access to those future steps, which were thus separable from the present issue.

To the powerless, however, no such identification could be made, because from their objective social position, consolidated in empirical historical experience, the processes by which the present step might or might not be converted into future elaborations were socially and intellectually impenetrable. From their social position it was therefore entirely logical to reject the equally logical, but contradictory definition of the issue by the elite, and to condense all future possibilities into the one present question. It was an undiscriminating, all-or-nothing stance, occasioned by their relationship to the axis of power.

VII CONCLUSIONS

My own suggested definition of technology is, of course, also political in that it highlights very different questions, and suggests different structures of naturalness and unnaturalness from other definitions, such as technology as 'tool', 'craft', 'package', or 'historical dynamo'. Although granting that technology does have intrinsic force and that this may well encompass and freeze, in its own way, the whole field of possibility for some societies or groups receiving a technology, the cultural process model does not commit the often-ensuing slide into technological determinism as a model of history. Nor does it encourage us to use such terms as 'technology' in an undifferentiated way, without attempting to understand people's different perceptions of control and responsibility in relation to it. Just as 'nature' acts as a mirror reflecting back our social and moral preoccupations, so too does 'technology'.[33]

I have tried to sketch a view of technology as a cultural process, attempting to link previous analysis of technology as social organisation with ideas about the way we structure experience of technology and its embedded social relations. By exploring the cognitive dimensions of these relations we may approach an understanding of the depth and complexity of the organisational dislocations which frequently beset modern technology. To see these as sociocultural allows us to conceive of them as rooted in cosmological commitments which the language of 'management' of 'organisational' difficulties may over-simplify. The pattern of possible cosmologies, their associated rationalities, metaphysics and individual identities and styles of interaction, are suggested by the grid–group cultural hypothesis.

The technological spectres, such as those I have discussed, act as a framework of interaction within and between these cultures. They also define these cultures by becoming central parts of their very identity. Sherry Turkle has discussed the fact that various technologies invoke strong personal feelings and intense relationships.[34] 'People develop intense and complex relationships with cars, motorbikes, pinball machines, stereos and ham radios.' Computers appear to have particularly strong properties in this direction. Turkle also recognised that such feelings can reflect external social and political concerns. However, what we are discussing here is not only relationships to, but *identification with* the technology, by fusion of personal or group identity with technological imagery. As we have seen, the purely mechanical technological metaphor can be re-animated by further metaphorical extension into images of intelligent controlling beings, but these are often alien, threatening and unpredictable, a metaphor for real social relationships.

This cultural process may occur on a microsocial scale in comparison to the overall organisational scale of the technology. Thus nuclear power station laggers are a small if crucial part of the overall system of nuclear power development and use. They install insulation at critical parts of the cooling circuit of reactors, so as to avert catastrophic thermal gradients and stresses which would crack the pipes and release radioactive gases. Their work is arduous and uncomfortable, involving the wearing of protective clothing in a maze of boilers and pipework. Interviews with laggers at the Heysham nuclear station in Lancashire, England[35] revealed that, well away from regular supervision as they are, they frequently remove gloves and dust masks to ease working conditions, even though the gloves are supposed to avoid possible corrosion from (acid) perspiration of the stainless steel pipes. When they need to urinate, instead of crawling laboriously back to an exit, and thence to the site W.C., they find a convenient corner on the job in the pipework system, releasing onto it a potentially corrosive liquid. When they lose a piece of equipment, they are supposed to report it at the end of the shift, and go back down with a supervisor to find it and 'sign it off'. Instead of subjecting themselves to an open-ended search in their own time, they quietly ignore and cover up the loss, thus leaving the equipment possibly to disrupt the highly sensitive, precision-flow dynamics of the cooling system when the reactor is started up.

Laggers are a culture unto themselves. They see the thing they are building as just a theatre for doing their work and drawing their pay. The identification they have with the technology is as with a white elephant – when asked to justify what looks like their potentially dangerous and irresponsible behaviour from the view of the nuclear technology as a whole, they do not see it as a piece of nuclear technology. They point in very well-informed fashion to management incompetence on a par at least with their own 'irresponsibility', and conclude that the technology will never come into being. Thus, seeing their behaviour as irresponsible in the overall technology context is, in their view, irrelevant. Arbitrary forces outside their control completely neutralise the implications of their own behaviour.

I would suggest that such cultures as fragments of overall technology systems are commonplace. As the technological division of labour becomes more elaborate and institutionalised, such groups become all the more segmented and isolated. In creating their own cosmologies out of this experience, they create a certain independence from the technolgy on which they depend. This 'independence' is, of course, not total, but gridded by the boundaries of related parts of the overall system. The growth of a quasi-independent cultural identity out of the corresponding social practices may stabilise the boundaries of activism of such groups by 'naturalising' the surrounding social 'landscape', to within limits that retain that dependence. However, this deeply ambivalent dependence may be misinterpreted as loyalty from the social distance of the axis of complacency, and the underlying alienation and cultural autonomy of such units never become apparent, except indirectly as technological (and maybe government) systems that do not work.

Notes

1. S. B. Barnes and S. A. Shapin (eds), *Natural Order* (London: Sage, 1979); M. Douglas, *Implicit Meanings* (London: Routledge & Kegan Paul, 1975).
2. M. Thompson, 'The cultural construction of nature and the natural destruction of culture', paper to International Conference on Nature,

102 **Technology as Cultural Process**

Culture, Technology, Stockholm, Sweden, September 1983; mimeo copy, IIASA, Laxenburg, Austria. A revised version of this paper is published as Chapter 4, 'Socially Viable Ideas of Nature: A Cultural Hypothesis'.

3. I have tried to make this connection through a detailed case study, B. Wynne, *Rationality and Ritual: The Windscale Inquiry and Nuclear Decisions in Britain*, British Society for the History of Science, Chalfont St Giles, Bucks, 1982.

4. B. Wynne, 'Redefining the issues of Risk and Social Acceptance: The social viability of technology', *Futures*, *15* (1983) pp. 13–32.

5. As offered, for example, by much mainstream risk perception literature. For a critique of this literature, see H. J. Otway and K. Thomas, 'Reflections on Risk Perception and Policy', *Risk Analysis*, vol. 2, *2* (1982) pp. 69–82. Another psychologist has observed that in risk analysis, 'we psychologists are a bit trapped by our own proficiency at being good experimentalists. We realize the importance of control and so we are drawn to those tasks in which we can exercise control. Hence our preoccupation with simple, static lotteries' [which are used as if they were real-life risk decisions], L. Lola, *Journal of Experimental Psychology: Human Perception and Performance*, *9* (1983) pp. 137–44. Also, Douglas MacLean, 'Is Rationality Extensional?', mimeo, University of Maryland, Department of Philosophy and Public Affairs.

6. As indeed in the long standing, excellent journal of that name, and in classical works in economic history, such as Carlo Cipolla, *Clocks and Culture 1500–1700* (London: Collins, 1967); Lynn White Jr, *Medieval Technology and Social Change* (Oxford: Clarendon Press, 1963).

7. David Edge, 'Technological Metaphor', in N. Wolfe and D. O. Edge (eds), *Meaning and Control*, Essays in the Social Aspects of Science and Technology (London: Tavistock Publications, 1973) pp. 31–64.

8. M. Douglas, *Social Factors in the Perception of Risk*, Report to the Russell Sage Foundation, New York, 1983; Douglas and A. Wildavsky, *Risk and Culture* (Berkeley: University of California Press, 1982); M. Thompson, 'Among the energy tribes', IIASA Working Paper, Laxenburg, Austria, WP–82–59. Published in *Policy Sciences*, 17 (1984).

9. See previous note. Also, G. Mars, *Cheats at Work: An Anthropology of Workplace Crime* (London: George Allen & Unwin, 1981); S. Henry (ed.), *Can I have it in Cash: A Study of Informal Institutions and Unorthodox Ways of Doing Things* (London: Astragal Books, 1981); M. Douglas (ed.), *Essays in the Sociology of Perception* (London: Routledge & Kegan Paul, 1981); S. Rayner, 'The Classification and Dynamics of Sectarian Organisations: Grid/Group Perspectives On the Far Left in Britain', PhD thesis, University College London, Department of Social Anthropology, 1979; M. Thompson, *Rubbish Theory* (Oxford University Press, 1979).

10. See, for example, Douglas, note 8, and Thompson's postscript to H. Kunreuther and J. Linnerooth (eds), *Risk Analysis and Decision Processes: The Siting of LEG Facilities in Four Countries* (Berlin: Springer Verlag, 1983).

11. See notes 2 and 10.
12. NIMBY, the 'Not In My Back Yard' theory of social protest, would have it that people only take interest in an issue when their immediate local interests, such as property, are threatened.
13. And the similar issue of political communication. See, for example, the discussions of 'communicative competence', in P. Connerton (ed.), *Critical Sociology* (Harmondsworth: Penguin Books, 1978).
14. Wynne, *Rationality and Ritual*.
15. For the concept of symbolic action, see, for example, M. Edelman, *Politics as Symbolic Action* (London: Academic Press, 1976).
16. See, for example, Douglas and Wildavsky, *Risk and Culture*. P. Lowe and J. Goyder, *Environmental Groups in Politics* (London: George Allen & Unwin, 1983), describes the strategic tensions for activist group leaders. I can also testify to this process from personal experience of such a (short-lived, but intense) role.
17. This is perhaps *the* fundamental issue in Western social science; see, for example, Douglas, *Social Factors*; Thompson, 'The cultural construction of nature'; Wynne, *Rationality and Ritual*, ch. 9; R. Unger, *Law in Modern Society* (London: Collier Macmillan, 1976).
18. The standard critiques of modern technocracy, such as J. Ellul, *The Technological Society* (New York: Vintage Books, 1964); H. Marcuse, *One-Dimensional Man* (New York: Beacon Press, 1966), tend to encourage such an exclusively monolithic view.
19. K. T. Erickson, *Everything in its Path: the Destruction of Community in the Buffalo Creek Flood* (New York: Simon & Schuster, 1976).
20. See also R. J. Lifton, *The Broken Connection* (New York: Simon & Schuster, 1979), for a psychiatric perspective adapting Freud's original concept of instincts and defence or 'blocking' mechanisms towards a more central role for images of life and death as explanatory factors for human attitudes and behaviour.
21. Erickson, *Everything in its Path*, p. 258.
22. Taken from *Financial Times*, London, Supplement on Atomic Power, October 1956.
23. R. Daly, 'The Specters of Technicism', *Psychiatry, 33*(4), 1970, pp. 417–31 (quote pp. 417, 421). For some earlier signs, see, for example, Roger Bastide, *Sociologie et Psychoanalyse* (Paris: Puf, 1950), describing the dreams of Indian tribes of automobiles breaking down; these are interpreted as a technical metaphor for sexual failure and derangement.
24. B. Bettelheim, 'Joey: A Mechanical Boy', *Scientific American*, March 1959, pp. 2–9. See also Edge's discussion, *Technological Metaphor*, pp. 50–2.
25. H. J. Otway and D. von Winterfeldt, 'Beyond Acceptable Risk', *Policy Sciences 14* (1982) pp. 27–45.
26. B. Wynne, 'Institutional mythologies and dual societies in the management of risk, in E. Ley and H. Kunreuther (eds), *The Risk Analysis Controversy* (Berlin: Springer Verlag, 1982) pp. 127–43.
27. Nor, of course, do the system's 'managers', but they too create elaborate, but different myths to fill in the rest in a way consistent with their

managerial position. It is just that these myths are different, even if they are more elaborate because they have more money and time spent on their articulation.

28. John McDermott, 'Technology: opiate of the intellectuals' in A. H. Teich (ed.), *Technology and Man's Future* (New York: St Martin's Press, 1974).
29. L. Winner, *Autonomous Technology* (Cambridge, Mass: MIT Press, 1978) pp. 33–5.
30. Ibid, p. 34.
31. M. Crozier, 'Les Développements Futurs de la Bureaucratie', *Courrier du Personnel, Commission of the European Communities, 416,* 29 July 1980, pp. 13–20.
32. See Wynne, *Rationality and Ritual.*
33. See, for example, the work of E. Wenk Jr, reported in *Futures, 15*(1) (1983) pp. 87–90, and discussed at a seminar at the International Institute for Applied Systems Analysis, Laxenburg, Austria, September 1983.
34. S. Turkle, 'Computers as Rohrschach: Subjectivity and social responsibility', in Bo Sundin (ed.), *Is the Computer a Tool?* (Stockholm: Almquist and Wiksell, 1981) pp. 81–99. Reprinted in *Transaction: Social Science and Modern Society, 17* (2) (1980).
35. During the research for a doctoral dissertation of Ian Welsh, to whom I am grateful for discussions around this point.

6 Water and the Flow of Power

Donald Worster

'Water taken in moderation cannot hurt anyone.'—Mark Twain

My subject is water and deserts. It is also power and domination. I want to use the desert environment to throw light on one of the oldest of human experiences: the social and political consequences that follow the process of ecological intensification. By that process I mean the effort to derive a greater economic return, or in some cases it may merely be a steady return, in the face of resource depletion, from any natural habitat. Throughout history ecological intensification, after a point, has forced societies into innovation; it has brought into play new tools, new techniques, new sources of capital, and new forms of social organisation to extract more of nature's wealth. My concern here is what this process of demanding more from the earth and of devising the means to get it does to the structure of power in a society. Water flowing through deserts can illuminate that universal process, I believe, in ways that no other environment can.

I do not mean, however, to posit a single, dogmatic answer to the question of what causes power to accumulate in a society. The ecological approach to that old puzzle offers only one of many plausible solutions, and it is not even to be demonstrated conclusively by any empirical methods. Glimpses of an important truth – that is all I mean to offer. Glimpses of a truth which has been overlooked in the standard analyses of politics, democracy, and technology. A common assumption, for example, is that a genuinely democratic society can flourish in a world where every desert has been conquered, where the earth is intensely managed on every hand, and where total dominion is the goal of humans in their dealings with nature. But is it really possible? Can democracy in fact thrive under such circumstances, or does it, along with nature, become a victim? Obviously what one means by democracy is important to such an inquiry. I do not take democracy to be merely a matter of elections and parliaments; I have in mind a deeper condition of widely distributed freedom and autonomy. In which communities, along with the individuals in them, retain considerable power to exercise cultural as well as economic and political self-management. Can it be assumed that democracy in this latter sense automatically follows in the wake of technological progress, or that democracy can exist apart from and in spite of the impact technology has on the natural order?

Take the desert then as a case study in the political, moral, and social issues posed by the human ecological situation. Let a river flow through that desert, bringing life to its apparent emptiness. Introduce a group of people intent on wresting a living from the place and see what the desert, the water, and the people do to each other. Beyond a threshold of modest resource development, I will argue, concentrated forms of power begin to emerge along the banks of the river and to rule over the people. Intensification of use eventually must give rise to potent anti-democratic forces, whatever their guise may be. The only remedy for that outcome, if people dislike it and demand something different, is to reverse the process of intensification, to 'liberate' the river and the desert wherever possible from domination. If I am right about that process, its outcome and its alternative, the desert example may help us understand the relationship between nature and society in the more dense, tangled environments in which most of us live.

For a long time now humans have been going to desert landscapes to free their minds from distractions, to sort out confusions, and to confront ultimate questions. There things can be seen with new clarity and force. The spareness of the desert concentrates the mind. Its scarcity narrows choices and makes them compelling. Consequences become unambiguous and long-lasting, like the bare bones of an animal lying preserved in the sand. The desert too is an environment that naturally provokes and tests theories, for theories are simplifications that work best in simplified places.

One of the most ambitious and influential ecological theories that has emerged from desert history is Karl Wittfogel's idea of hydraulic society. It is essentially a theory of power. The command of water in arid environments, Wittfogel argued, gives rise to new agglomerations of power or enhances old ones. He was particularly interested in explaining the stagnation, as he viewed it, of the major Asian civilisations over thousands of years. The problem of explaining why those civilisations failed to maintain their lead in social evolution, why they were eclipsed in modern times by the Europeans, first began to be considered in eighteenth-century England, which was amazed by its own rapidity of change and newly aware of the past grandeur of the Orient; from there the problem of a stagnant Asia passed down to Karl Marx and on to Wittfogel. The answer given by Wittfogel, beginning in the 1920s, was an ecological one: in their struggle for existence in arid environments the Asians, he pointed out, had developed massive, sophisticated systems of irrigation, which in turn had brought into being highly centralised structures of political power governing large but impoverished peasant populations. He called them the 'hydraulic societies'. The elaborate apparatus of water control, he maintained, and the managerial organisation needed to operate it became, after a long period of intensification, a technological and institutional strait-jacket preventing radical change. In countries like China, India, Mesopotamia, and Egypt, preserving the irrigation system intact was the first priority. The system was vulnerable to a thousand mishaps; highly regimented control was the price of survival. Wittfogel went on to describe those societies as 'despotic' and 'totalitarian', even going so far as to explain the modern totalitarianism of the Soviet Union as an 'Oriental legacy'.[1]

Wittfogel's critics over the past several decades have had a field day with his theory, charging that it is too sweeping, ethnocentric, and simplistic, even that it is dead wrong. In many particulars they have been right. Nevertheless, a number of ecologically oriented anthropologists have continued to follow Wittfogel's lead, though in more restrained fashion. So has Lewis Mumford, whose two-volume work, *The Myth of the Machine*, compares the modern military–industrial complex in the United States with the ancient hydraulic societies.[2] Mumford has, in his inimitable way, extracted the most useful and credible argument in Wittfogel's Asian studies and applied it, as Wittfogel would not, to the twentieth-century and the Occident. That argument, simply put, is that the domination of nature leads inescapably to the domination of some people by others. Or, to translate the argument into the terms of our case study, the technological conquest of a river flowing through a desert transforms those specific individuals doing the conquering into a power elite.

That insight – or hypothesis, as one might say – was not Wittfogel's alone. He shared it with his associates in the Institute of Social Research, established at Frankfurt, Germany, in 1923. What Wittfogel did was apply the insight to the archaic Asian irrigation regimes – and thereby restrict its use to a single and remote social type in ancient history. The broader and contemporary implications of the idea were left to be developed by others in the Institute, in particular, by the Institute's two leading lights, Max Horkheimer and Theodor Adorno, and later on by Herbert Marcuse. Most of their ideas are well known today; their shock and revulsion at the sudden emergence of Hitler, Fascism, and other forms of totalitarianism in civilised cultures; their interest in the Freudian psychology of repression; their critique of mass society; and their rejection of positivistic science. Less appreciated, at least until rather recently, has been the central place nature occupied in their thinking, especially in Horkheimer's. As Martin Jay puts it, the Institute replaced class conflict with 'a new motor of history . . . the larger conflict between man and nature both without and within'.[3] In fact, it is accurate, I think, to describe the Institute's work as, *au fond*, the formulation of an ecological theory of history. The focus of that theory is not so much the impact nature has had on society as the transformation humans have worked on themselves through their working on nature. When the age-old struggle for human existence within nature became the modern struggle for human domination over nature, the Frankfurt theory goes, history began a tragic decline into unfreedom and barbarism.

Modern societies, from the eighteenth-century on, the Frankfurt philosophers argued, have been devoted to a project of total domination of nature. Though originating in western Europe, the project has, by the twentieth-century, become transcultural in scope. It has its roots mainly in capitalism and market-place thinking, but it has spread into and deeply affected the socialism of Marx, Lenin, and Mao; today in fact it is the central project of advanced industrial culture wherever found or pursued. It promises a more democratic future for the human species and an expansion of individual freedom, but so far the result has been precisely the opposite.

What then is this project of domination? Horkheimer and Adorno understood it to consist of both an intellectual and a technological element. In the first place domination involves 'the disenchantment of the world'. Nature is to be emptied of all intrinsic meaning and value, all beauty and mystery, to be 'degraded to mere material, mere stuff to be dominated, without any other purpose than that of this very domination'.[4] Then the disenchanted world, no longer alive but dead, is to be brought under the absolute control of humans through instrumental or technological rationality. Nothing in nature is to elude mastery; nothing that is potentially useful in it must go unused. Thus the project aims to go far beyond giving people a merely adequate means to survive on the earth, for that would entail a fixed list of specific needs to be filled. Domination aims rather at setting up nothing less than a totalitarian government over nature. It seeks to transform the non-human environment into a monolithic unity ruled over by rigid, authoritarian methods. The excercise of that command becomes in the project an end in itself – an end that has no ending, no point of closure, no sense of limits.

Under this project of total domination, the individual's power over the natural environment (power realised in the form of wealth, comfort, and gratification) increases remarkably. but, the Frankfurt circle repeatedly emphasised, some people gain far more power than others. They may be capitalists, scientists and engineers, or bureaucrats presiding over a state agency – they are whoever is directly involved in and responsible for the conquest of nature, intellectually or technologically. As the project progresses, they become more and more indispensable. They control public policy, they constantly devise new technologies to remedy the imperfections of the old, and they hold the threads of life and death in their hands. They come to manage not only nature but human society as well. 'The human being', Horkheimer wrote in *The Eclipse of Reason*, 'in the process of his emancipation, shares the fate of the rest of his world. Domination of nature involves domination of man'.[5]

For people who have lived long in the midst of the megalopolis, seldom if ever escaping from its technological envelope, this ecological theory of the Frankfurt philosophers may seem implausible. To be accepted in a more than academic sense, it requires a capacity to see oneself in the mirror of nature and to say spontaneously, 'there I am and I am nature'. When, however, people are unaccustomed to discerning in that mirror their own face – their own fate – they turn to other explanations. They may indeed feel a loss of autonomy in their lives, they may complain about the 'corporations' and the 'bureaucracy', may even rage against their personal and community powerlessness. But the ultimate source of the problem appears to rest within society, in social organisation, not in the project of environmental domination.

That is why a desert landscape and its history can be peculiarly instructive. We can find there the project revealed in the clearest light around. We still may not see, of course, or may not accept what we see, but the choices confronting us in the desert are harder to overlook: either we must knowingly accept the project and the human consequences it entails, or we must reject it and search for a different relationship with nature.

The Colorado River basin in the southwestern United States affords perhaps the best opportunity we have anywhere on the planet for an inquiry into the project of domination in modern history and its social ramifications. Most of the basin is an intensely arid landscape where a little water is appreciated by living creatures like a transfusion of blood by a dying man. Not far from the river, among its many appendages, is Imperial Valley, which, as I will elaborate in a while, has become, in the space of less than a century, an intensely irrigated environment and the home of an agribusiness complex that exercises an increasingly global reach. Horkheimer, Adorno, and Marcuse all lived for a number of years within a few hours' drive of the basin and the valley, apparently without realising how well they illustrated and extended some of their ideas. Karl Wittfogel also came to live on the west coast of America, on the edge of what has become the most elaborate hydraulic society in history, and even more surprisingly than Horkheimer and the rest, he never paid it much attention.[6] We should undertake to remedy their oversight and learn something useful about the ecology of power in this American desert place.

The Colorado was one of the very last major rivers to be discovered and explored by civilisation. In 1869, Major John Wesley Powell and nine other men shoved off from Green river, Wyoming, to ride the unknown river through some of the deepest canyons and most dangerous rapids on earth. Three months later two of the original four boats, carrying six of the ten men, pulled in at Grand Wash Cliffs, a hundred miles downstream from the Grand Canyon. There-after, Powell went into the federal bureaucracy and gave his best energies to the project of river domination. 'All the waters of the arid lands', he predicted, 'will eventually be taken from their natural channels.'[7] Even the mighty Colorado would be turned out of its course in order to irrigate the desert, redeeming it, transforming it from a condition of waste into a garden of affluence. 'The greatest possible development', 'total utilization', were phrases he used to describe the project of domination. What he saw in the river were mainly quantities that could be measured and augmented: so many acre-feet of water, waiting to be captured and forced to provide an income, a potential flow of cash and technological power. What he did not acknowledge, or did not care about, was what others have described as the flow of life in that water, the mystery in it that surpasses understanding.

Today the project of river domination is virtually finished in the Colorado basin. More than thirty dams have been built on the main river and its tributaries. An immense amount of water is steadily pumped out of reservoirs there to support farming on the Great Plains, in central Utah, in Arizona, in southernmost California, and in Mexico; and urban people, swarming into the desert cities of Denver, Las Vegas, Phoenix, Tucson, Mexicali, and Los Angeles, all suck water out of the river too, or expect to do so some day soon. The effect of this intensified use has been devastating for the natural Colorado. For the past twenty years it has not in normal seasons reached the sea. Its lower reaches are now an artificial drainage ditch, dredged and lined, sporadically carrying the heavily saline run-off from lettuce and cotton fields. The Colorado, one recent writer has said, is 'a river no more'.[8] Nowhere has the disenchantment of nature and the triumph of technological reason been more dramatic and thorough than in this corner of the earth.

All this was more or less in the mind of John Wesley Powell a hundred years ago. What he did not foresee, perhaps could not have foreseen in his time, was the full process of social reorganisation required to carry out such domination. Powell naively believed that total domination could be accomplished by small communities of farmers working largely with their own tools and money up and down the river, each community existing largely independent of the others, each free of outside interference in its affairs: by a series of democratic 'commonwealths', as he called them. Instead, the Colorado project has created in the basin, or at least has promoted there, a highly centralised, bureaucratised political order; a technocratic corps of river experts; a set of corporatised agricultural entities with near total dominance over rural affairs; and a class system in which there are sharp disparities of wealth and influence.

Those outcomes are most starkly manifested in Imperial Valley, which lies west of the Colorado River and directly north of the United States–Mexico border. In 1900 this valley was still known as the Colorado Desert. Once an extension of the ocean, it was cut off a few million years ago by the river's delta and then dried up, forming a vast bowl, whose deepest point lay several hundred feet below sea level, a bowl of intense heat and virtually no rain, one of the most formidable environments in North America. The great river ran high up along one edge of the bowl. To bring irrigation water into the valley required simply cutting a notch in that edge and letting the river run downhill. A corporation formed in 1892 to do just that, but it was not until 1901 that the first settlers arrived to begin farming.[9]

According to the most recent tabulations, Imperial now ranks as one of the four richest counties in the United States in total agricultural production. Its average farm is worth well over $1 million. Half the acreage is owned by wealthy non-residents, including several multinational corporations. A few of the long-time resident farmers have begun to throw a large shadow across world food markets; among them is Bud Antle, now incorporated, who has expanded his operation into Africa, growing vegetables there for export to Europe. Yet, along with its rural moguls, Imperial Valley includes many poor inhabitants too. Due to the fact that a large part of the population (12000 of them, compared to less than 700 farmers) are seasonal labourers, usually with Hispanic surnames, working on the factory farms, the area has one of the lowest income averages in California.[10]

Command of the Colorado River explains the valley's extraordinary rise from unsettled desert to luxuriant wealth. It also explains in large part the social structure that has evolved in Imperial. Irrigated farming depends on high-return-per-acre crops to be profitable. Those crops are typically labour intensive ones; to get planted and harvested they require a great deal of stooping and picking by an army of hired workers. Crops that are not gathered at exactly the right moment spoil quickly, so the valley's farmers have had a strong incentive (and not a little government sympathy and help) to prevent unionisation of those workers. A chain of social consequences, in other words, has followed the project of domination. And it does not stop there.

The total management of a mobile, elusive, dangerous river, carrying millions of tons of silt downstream every year, rising and falling dramatically with melting snow and summer flash floods; the construction and maintenance of a labyrinth of canals and drains; the capitalisation of headworks, dams, siphons, and pumping stations – all this obviously could not have been done by ordinary farmers loosely associated in a traditional rural culture. What was needed in Imperial, and what has evolved there, is a complex power centre to do that work. Settlers organised themselves in 1911 into an entity called the Imperial Irrigation District. Ever since then it has performed as a kind of modern business corporation, selling bonds to finance its programme, hiring engineering talent, and welding disparate farmers into a single unified institution. The most recent *Organization*

Manual published by the district is a massive document, taking hundreds of pages to trace the lines of authority that run from the Board of Directors (elected by farmers acting as shareholders) through managers, economists, data processors, hydraulic engineers, auditors, and so forth. In his study of the district's political evolution, Ernest Leonard describes it as 'a conservative, paternalistic, and self-protective managerial system' that has become immune to the diverse needs in the community. It is effectively run, he writes, by an elite group in the valley, and year after year there is little challenge or alternative to its reign.[11]

The chain of consequences goes on. To intensify their use of the Colorado, farmers in Imperial Valley have been forced to turn to a succession of outside agencies, each one more powerful than the last. In the beginning they relied on several private corporations, including one of the largest railroad empires in the country, the Southern Pacific under E. H. Harriman, for loans, technical assistance, and direction. Without that help, the river, threatening again and again to break through its edge and flood the bowl, would have defeated them.[12] Since the 1920s the valley's farmers have relied on the federal government, specifically the Bureau of Reclamation, the largest irrigation bureaucracy ever assembled in history. BuRec, as it is called, is famous the world over for its ambitious water projects: its slogan, blazoned on report and book covers, is 'total use for greater wealth'. Today the agency supplies water at very low rates (due to the fact that they are subsidised by taxpayers across the nation) to farmers all over the American West, and in return it asks only for a share of the administrative power. The Imperial grower accepts the federal water, lives better than ever, and becomes a ward of the distant government. If for some reason, BuRec could no longer effectively control its far-flung apparatus, if the ecological problems associated with intensive irrigation became unmanageable, or if the central government began to favour different groups, say, the industrialists in Los Angeles, then the Imperial Irrigation District would be left dry and impoverished. Its own power thus depends on water flowing down from a power higher up.

Who then really dominates society in the American desert environment? There is no simple answer. The power elite in this modern hydraulic society does not rise to a single pyramid point, as Wittfogel found in ancient Egypt. Rather, a number of interlocking hierarchies rule, forming an alliance based on money, class, and expertise, one that merges private interests and state authority.

Whatever their differences from time to time may be, they are united in the project of environmental domination. Their power, they are aware, stems from their shared role in that domination. Left out of the alliance are those whose contribution is merely to pick a basket of hops or carrots, as are those who merely sit by the river and contemplate its flow from dam to dam. Even those who circle endlessly in their motorboats on the stilled river, enjoying the leisure provided by BuRec recreation planners, are not part of the power structure. They live, as Horkheimer and Adorno have helped us see, 'in the world of the administered life'.[13]

The Colorado story has many parallels elsewhere, including along the rivers Volga, Zambezi, Nile, and Indus. Though more completely achieved than those other efforts at water control, the Colorado project is not significantly different from them in its purposes. In some places 'people's commissars' instead of agribusinessmen may divide the water or direct the farming; or workers rather than a handful of private interests may be said to 'own' the land and water with which they work. In some basins the wealth made possible by domination may be better distributed than in others, or the power elite may be more or less benevolent – and those are not trivial differences. But in so far as the democratic qualities of freedom and self-management are concerned, the distinctions from river valley to river valley are increasingly unimportant. Their organisation manuals are indistinguishable. In all of them the 'administered life' is coming to be the common experience for rivers, deserts, and people alike.

Thousands of tourists gather every year on the crests of Hoover and Glen Canyon Dams, looking down on the governed Colorado River hundreds of feet below, admiring the grandeur of the curving concrete wall on which they stand, talking of the beauty of blue water lapping quietly against brown rock. Down in the bowels of Glen Canyon Dam immense dynamos hum, generating electricity for distant cities, while an electronic toteboard tells the visitor how many dollars are being made from power sales instant by instant. But nowhere in the visitor centres and their rituals of celebration is this question raised: what comes next, now that the project of domination here is complete? There are at least three answers to that question, three possible futures, which have been suggested by scholars and philosophers. I now want to examine each of them.

One set of possibilities for the future of the American hydraulic society emerges from study of its ancient counterparts. Marvin Harris, for instance, following Wittfogel's lead, argues that the old versions went through a long period of social stagnation, during which the power elite remained unchanged in form but was populated by a succession of dynasties. Harris calls this phenomenon the 'hydraulic trap'. Egypt may be the best historical example of this outcome, at least up until its invasion by the British and French imperialists in the nineteenth century. In other cases, however, the stagnation came abruptly to an end when ecological problems became unsolveable. Mesopotamia and Mohenjo-Daro are cases in point: silt accumulated in their waterways and salt built up in their fields from intensive irrigation until the effort to maintain their systems became too great a burden, and they let them collapse. There is in both sorts of fate a rather grim lesson to be learned from the archaic desert regimes, one not calculated to amuse or inspire the dam tourists.[14]

A second set of possibilities can be derived from the highly disparate writings of the Frankfurt school. Aside from Wittfogel, they did not address explicitly the subject of water control as a form of domination, nor did they make any specific recommendations about the next stage in the Colorado's history. But they did have much to say about where the human-nature relationship might go, and should go, in the future. They would being, one and all, with a repudiation of the project of domination, at least for people living in the advanced industrial societies for whom they wrote. What that repudiation would mean in practice, however, is not altogether clear; and it is easier to discern what it would offer for humans than nature. No one in the Frankfurt group was quite willing to give up the economic bounty made possible by domination. What they appear to have wanted was a halt to the project's expansion rather than its deconstruction.

Herbert Marcuse's writings offer the most systematic, coherent vision of a future beyond domination. It is a future in which humans – all humans, not merely those of the privileged classes – are free at last from the burdens of being dominators. Those burdens, Marcuse argues, have been repressive and distorting. Described in the most familiar terms, they are the burdens associated with the traditional work ethic: postponement of gratification, labour without joy, a rigid control of one's libidinal energies in the interest of production and accumulation. At an earlier point in history, Marcuse goes on, when the human condition was one of deprivation, the drive for environmental domination made sense, and its price in psychic distortion simply had to be paid. 'Society must first create the material prerequisites of freedom for all its members before it can be a free society', he writes; 'it must first *create* the wealth before being able to *distribute* it according to the freely developing needs of the individual'. There must be, in other words, a 'conquest of scarcity'.[15] The crucial issue is knowing when that conquest has been satisfactorily completed. For Marcuse, the point has long arrived, and it is now time to begin relaxing in the freedom made possible by the conquest. People should henceforth work only at what is truly satisfying to them. They should ease the tight control over their feelings and natural instincts; they should free their minds and bodies from antiquated bourgeois attitudes, remove the dams from their inner rivers and let the pent-up waters flow.

A new human project then should replace domination, the project of rational gratification. It would have, Marcuse maintains, radical political effects, for it would destroy the power elites and lead to a democratisation of self-fulfilment. Instead of being driven and manipulated by the discredited elites to keep the assembly lines rolling, men and women would take command of their own lives. But this new project and the democracy it allowed depended, Marcuse insists, on maintaining some technological control over non-human nature. It was his belief that automation would be the basis for a fundamental change toward a civilisation oriented towards rational gratification. 'Complete automation in the realm of necessity would open the dimension of free time as the one in which man's private *and* societal existence would constitute itself. This would be the historical transcendence toward a new civilisation.'[16] In other words, people would be relieved of all the chores of domination; nature, on the other hand, would be freed only from the pressure to fill an unending list of consumer demands.

What Marcuse failed to realise in his Utopian musings was that automation, even for a limited list of 'necessities', must require a maintenance crew, along with an organisation manual, clockwork regimentation, and a measure of continuing control over those set free to play. How much change then would his Utopia require in the existing hydraulic regime along the Colorado River? A precise answer is not easy to come by, for it depends on how much water Marcuse thought we should want to drink – that is to say, how much consumption was assumed in his idea of 'vital need'. One thing is clear: he was firmly opposed to the practice of 'asceticism'.

To illustrate the Frankfurt project of democratised gratification (and I think this image applies to Horkheimer and Adorno as much as Marcuse) we can picture a green oasis, where the date trees are filled with plump fruit, where the senses are filled with the fragrance of spice and wine, where every person is his own master, and where the water drips unceasingly through the automated plumbing. Omar Khayyam joined to Marx and Freud. Marcuse emphatically did not want to live in a desert. He preferred life in an oasis. The Colorado River consequently would still be dammed and siphoned off to provide a basis for his new stage of civilisation.[17]

The vision of an oasis appearing out of the desert is an old, old one in history, though in the hands of the Frankfurt philosophers it would assume a radical content. How is it possible, Marcuse asks us, to achieve a reconciliation with nature until after the survival of each of us is assured? And does that mutual survival not require a technological conquest – making the reconcilation, ironically, the product of domination? It is much easier, he would say, to achieve both social justice and ecological harmony in an oasis than in a desert, in a state of abundance rather than want. There is, however, a persistent uncertainty plaguing that reasoning: can Marcuse deliver what it says he will, a democracy of freedom, self-management, and autonomy? Can a democracy be reliably based, as he would have it, on an appeal to gratification? Always, it seems to me, there must be a flow of power into the hands of those who provide the technological basis for gratification, no matter how strictly it may be limited in material terms. Moreover, it is extremely difficult to locate that point where physical gratification begins to transcend itself, where it becomes the pursuit of spiritual and intellectual pleasure, as Marcuse expected it to become. Who is to say that Las Vegas is not the oasis in which people really want to live?

The mention of America's favourite fun city is a reminder of how far the desert has already been transformed into an oasis of gratification. Las Vegas sits only a few miles from the Colorado River, its brilliant, fantastic neon advertising lit by power from Hoover Dam. Water bubbles out of fountains in an endless stream, though the city is surrounded by an intensely arid region. Its night is abolished by electricity, its heat by air conditioning; food is cheap, work is forgotten. And this city dedicated to the pleasure principle is, as everyone knows, organised and run by powerful crime syndicates, politicians, and the mass entertainment industry. That assuredly is not at all what Marcuse had in mind for his Utopia of liberation, for his '"garden" which can grow while making human beings grow'.[18] Yet the distinction between the two oases is not at all easy to draw.

Hedonism has been a notoriously untrustworthy guide in many

matters. It has had a way of muddling thought, undermining independence, and betraying its proponents. And that finally is where the
Frankfurt philosophers leave us, muddled and too easily betrayed.
We must forsake the old bourgeois drive to master and *accumulate*,
they tell us; we must now seek to master and *enjoy*. New 'projects'
take the place of the old ones – 'the abolition of toil, the amelioration
of the environment, the conquest of disease and decay, the creation
of luxury', is the list Marcuse gives – and all of them require
considerable manipulation of nature.[19] The earth is supposed to be
freed at last from the project of domination, but it is also now to be
'pacified' and turned into a garden of delights. Machines are to
release the farm workers from all their hardship in Imperial Valley
and, unaided and untended, grow the grain and fruit the world wants.
But can the automated machines and dams and irrigation canals also
be counted on to grow democracy and freedom? Or are they more
likely, in the light of previous human experience, to grow entrenched
power and all-night casinos?

The Frankfurt dream of a paradise beyond domination, however
appealing and persuasive it may be, ends in the wrong landscape.
Where it should end, where it must end if it is to succeed as a
revolutionary vision, is in the desert. The qualities of the desert are
not attractive to the hedonist, whether radical or conservative in
politics, for they represent scarcity and deprivation. But there is a
more positive way to regard the desert, and it is therein we can find
a third set of possibilities for the hydraulic society's (and by extension,
the advanced industrial society's) future.

Deserts are landscapes where people have had to learn to do
without. They are the quintessential nature that the project of
domination has been most eager, for more than two hundred years
now, not to mention Wittfogel's Asian despotisms, to overcome and
replace. They are the hot, dusty lands where only small, roving bands
of people can find enough to eat. They are the cold dry steppes of
northern latitudes and continental interiors, the low coral atolls
scattered at sea. In all those deserts life can certainly be found, but
it gets along only by learning the virtues of austerity and humility.
Despite these constraints, humans have produced some of their finest
achievements in the desert: many of its loftiest moral and religious
ideals, moving examples of self-sacrifice, noble works of art, strong
communal bonds, and extraordinary personal competence. Desert
peoples, in contrast to those in oases and irrigated river valleys, have

been hard to subdue, as T. E. Lawrence understood some decades back, even by modern forces armed with sophisticated weapons. Can an age that feels itself overly managed neglect then to preserve and cherish its deserts? Are we well advised to continue taking out the water from our rivers to make more oases and gardens? Is oasis life really so appealing after all?

A number of recent signs may indicate that a shift toward 'desert thinking' is beginning in the more affluent cultures. Consider, for example, that the Americans, along with several other peoples, have set aside large desert areas for camping and contemplation. Such behaviour required an astonishing turnaround in some old cultural attitudes; it signifies a new thirst for being thirsty, a need to leave behind (if only for a while) the clutter of advanced industrial society and the easy gratifications it affords in order to live a little closer to the bone. It is related, I believe, to a spreading interest in reordering one's life around the principle of material simplicity. According to a recent report by the Stanford Research Institute, some five million Americans, or three per cent of the adult population, now practise an economically simplified life-style.[20] That estimate may be too large, but unquestionably there is a group of people, increasing in country after country, who are making a determined effort to free themselves and their communities from consumer gratification as a basis for living. Although they may not in every case be actually seeking out deserts for their inspiration, they are, in a sense, a kind of desert-affirming people. They want neither a made-over, lushly producing Imperial Valley, a high-rolling, high-spending Las Vegas, nor Herbert Marcuse's dream oasis of libidinal release.

The recent turning toward deserts and toward simplicity of consumption departs from both the bourgeois emphasis on postponing gratification and the Frankfurt emphasis on escaping repression. The new point of view is that discipline and restraint are not necessarily undesirable, nor are they to be automatically associated with the project of environmental domination. On the contrary, self-discipline may be regarded as the only true antithesis of domination. Liberating nature from the threat of endless conquest, endless intensification of use, the argument for simplicity goes, is not likely to be achieved under any social philosophy based on hedonism. Instead, it will require a cultural dedication to the mastery of self. That does not have to be only a private strategy; conceivably, it could have a public, political dimension too, including the decentralisation of production into local and regional modes, the development of new forms of technology that interfere less with natural processes, and the setting of personal income ceilings and the redistribution of the surplus.

Despite these evidences that people have begun to rediscover the moral ideal of self-imposed restraint, the obstacles to a wholesale move in that direction remain very imposing. It may well turn out that the human appetite will not be, cannot be, moderated. On a planet teeming with four or eight or twelve billion people, examples of restraint may remain what they are now: a series of minor, isolated gestures, unthinkable to the starving, unacceptable to the aspiring, unappealing to the affluent. If that is to be the case, then the fate of the Colorado River will inevitably become the fate of every river on earth. Their waters will be turned out of their channels onto every remaining scrap of wasteland, changing the last deserts into food, fibre, and energy factories. Whether that future arrives by choice or by necessity is immaterial; in either case, I'm afraid, democracy – in the sense of freedom from centralised authorities and oppressive hierarchies, in the sense of escape from the managed life, in the sense of ordinary people exercising a high degree of autonomy, cultural and political, in their lives – will not remain a realistic social ideal. Where wants and needs are out of self-control, where they cannot be defined or filled by the person and the immediate community, power must gravitate farther and farther away.

Notes

1. Wittfogel's major work was *Oriental Despotism: A Comparative Study of Total Power* (New Haven: Yale University Press, 1957). The phrase 'hydraulic society' first appeared in his article, 'Die Theorie der orientalischen Gesellschaft' (1938).
2. Mumford argues that the Age of the Pyramids has returned in spirit and ethos, bent as before on establishing 'absolute centralized control over both nature and man.' See *Technics and Human Development* (New York: Harcourt Brace Jovanovich, 1967) p. 207.
3. Martin Jay, *The Dialectical Imagination: A History of the Frankfurt School and the Institute of Social Research, 1923–1950* (Boston: Little, Brown & Co., 1973) p. 256.
4. Max Horkheimer, *The Eclipse of Reason* (1947; New York: Continuum, 1974) p. 97.
5. Ibid., p. 93.
6. Horkheimer and Adorno lived in exile in Pacific Palisades, California,

from 1941 to 1949. Wittfogel became an American citizen and taught for many years at the University of Washington in Seattle, while Marcuse spent the last part of his career in San Diego, California.

7. J. W. Powell, *Report on the Lands of the Arid Region of the United States* (1879; Cambridge, Mass: Belknap, 1962) p. 54.
8. Philip Fradkin, *A River No More: The Colorado River and the West* (New York: Knopf, Alfred A., Inc., 1981) p. 16.
9. There is no general history, but see Helen Hosmer, 'Triumph and Failure in the Imperial Valley', in T. H. Watkins (ed.), *The Grand Colorado* (Palo Alto: 1969) pp. 205–21.
10. Paul Barnett, *Imperial Valley: The Land of Sun and Subsidies* (Davis, Calif: 1978) pp. 3, 60.
11. Ernest Leonard, 'The Imperial Irrigation District: Agency Behavior in a Political Environment' (PhD thesis, Claremont Graduate School, 1972) pp. 8–9.
12. The story of Harriman's battle with the river is told by George Kennan in *E. H. Harriman: A Biography* (Boston: 1922), vol. II, 136–73.
13. Max Horkheimer and Theodor Adorno, *Dialectic of Enlightenment*, trans. John Cumming (1944; New York: Continuum, 1972) p. ix.
14. Marvin Harris, *Cannibals and Kings: The Origins of Cultures* (New York: 1977), ch. 13. Thorkild Jacobsen and Robert Adams, 'Salt and Silt in Ancient Mesopotamian Agriculture,' *Science* 128 (21 November 1958) pp. 254–8.
15. Herbert Marcuse, *One-Dimensional Man: Studies in the Ideology of Advanced Industrial Society* (Boston: Bevan Press, 1964) p. 40.
16. Ibid., p. 37.
17. Marcuse differentiated his 'mastery' of nature from that of industrial society as a 'liberation' and a 'pacification' rather than a 'repression'. Nature would cease to be 'mere Nature' as its violence, misery, and cruelty were reduced. The lion would lie down with the lamb, with Marcuse in the middle. See *One-Dimensional Man*, p. 236.
18. Marcuse, *Eros and Civilisation: A Philosophical Inquiry into Freud* (New York: Vintage, 1955) p. 197.
19. Ibid., p. 193.
20. Stanford Research Institute, *Business Intelligence Program*, no. 1004 (1976). See also Michael Phillips, 'SRI Is Wrong About Voluntary Simplicity', *CoEvolution Quarterly* (Summer 1977) pp. 32–4.

7 Conserving Nature and Antiquity*

David Lowenthal

The heritages of the natural and of the man-made environment exhibit remarkable parallels along with instructive differences. Campaigns to conserve nature and to preserve remnants of antiquity have become increasingly linked in content and in opposing forces. This essay traces the careers of these two campaigns, explains how and why they converge or diverge, and offers a critique of their rationales and effectiveness.

Many today, beset by momentous and often cataclysmic change, lament the degradation or disappearance of familiar locales and landmarks. Because the loss of habitual environments and traditional milieus threatens our very sense of identity, we treasure their surviving vestiges all the more. A deeply felt need for tangible relics of both nature and culture fuels crusades to conserve and preserve. Relics and landscape features regarded as symbolically important vary widely with individual, culture, and ideology, but the attachments they reflect are none the less universal. They are expressed by peoples at all levels of technology and of every political persuasion.

I COMMON ASPECTS OF NATURAL AND CULTURAL HERITAGE

The World Heritage Convention gives global sanction to the conjunction of nature and culture. Sites embracing the world's paramount treasures of environmental and human history receive similar acclaim and attention. Many national organisations affirm the same conjunction. For example, the Caribbean Conservation Association

* I am grateful to Colin Clarke and Oxford School of Geography students, to Robert Z. Melnick and his colleagues and students at the University of Oregon, to Bryn Green of Wye Agricultural College, and to my own Conservation and Preservation students at University College London for inspiring my exploration of the conjunctures and divergences between these topics. Some of these themes are discussed in another context in my 'Heritage and Its Interpreters', *Heritage Australia*, 5:2 (1986).

attends alike to coral reefs and country houses, to mansions and mangrove swamps, to tropical rain forests and ruins of sugar factories.[1] Britain's National Heritage Memorial Fund helps to save for the nation the Communion breadholder of Mary Queen of Scots and the Greater Horseshoe bat, a fossiliferous chalk cliff and a crumbling castle.[2]

Beyond the umbrella term 'heritage', however, what have these residues of nature and culture in common? What has the Grand Canyon of the Colorado to do with Grand Central Terminal in New York, the Sussex Downs with St Paul's Cathedral, the Somerset Levels with Somerset House?

Similar motives of preservation

One common feature is the reasons advanced for saving them from extinction. Those who strive to preserve facets of the natural and the man-made environment often do so for the same reasons. Remnants of nature and culture are both felt to provide identity and guidance, to enhance life through their rich diversity, and to benefit long-term personal and group interests. The likeness of their most zealous supporters underscores this affinity. The same kind of elite cadre – largely well-born, well-off, well-educated – leads both crusades. Indeed, many figure prominently in both realms of concern. As I show later, the birth of seminal ideas about man's relations with the built and with the natural environment exhibit the same convergence.

As modern attractions, nature and antiquity have substantially similar appeal. Natural wonders and historic splendours are often advertised and presented in much the same way. In many lands they have alike become essential sources of employment and revenue, particularly of foreign exchange.

Similar risks of loss

Aspects of nature and antiquity both warrant protection because they are non-renewable and in limited supply. Once gone, they are gone forever. To be sure, living things are capable of reproducing themselves, and new artifacts, some of which will in time become relics, are constantly created. But a species, a gene pool, an ecosystem are neither more nor less capable of regeneration than a cathedral, an earthwork, or an Old Master painting. They vanish at varying tempos, but none endures forever.

Both the natural and the man-made environment embody tangible memories that crucially connect us to the world we live in. The loss of either kind of environmental root imperils our sense of identity. Threatening both the natural and the cultural heritage are the same formidable forces: development, productivity, innovation, the cult of creativity, and technology. The bulldozer symbolises the agencies that transform the world apace at the cost of both aspects of heritage. Modern technology ever more drastically alters the artifactual and the natural environment, threatening to obliterate all of it along with ourselves. The hubris of developers, the fury of iconoclasts, the destructive greed of collectors may be no greater than in the past, but the power now at their disposal magnifies the likelihood of loss whether the target be a rare species, a prehistoric site, or an old church.

II DIVERGENT CONCERNS FOR NATURE AND FOR CULTURE

In certain other respects, however, the heritages of nature and culture seem dissimilar, and these differences, along with the resemblances, shape the efforts made to safeguard them.

Perceived differences between living assemblages and inanimate artifacts

We generally accept that by their very nature living beings perish within relatively brief lifespans that can scarcely be prolonged; it is a futile exercise to preserve any particular plant or animal. We also know that particular churches, paintings, and other precious artifacts likewise perish, but usually at slower tempos that outlast human lifespans; and their existence is capable of indefinite extension through conservation.

Hence nature conservation mainly focuses not on individuals but on groups – less on any particular tree or bird or whale than on species or ecosystems, aggregates of creatures. (There are exceptions: for example, the British Site of Special Scientific Interest (SSSI), small in area and unique in character, has some of the individuality of a cathedral or a country house.) Preservers of antiquity, by contrast, tend to focus on discrete relics. To be sure, concern with neighbourhoods, districts, entire towns, whole categories of structures, is greater than it was; over and above any single church, we value medieval parish churches as a 'species' for the character they lend the English village. But areas and aggregates remain marginal to historic preservation's primary emphasis on individual structures.

In dealing with the realm of artifacts, we usually notice and identify particular relics; we rarely do this with particular plants or animals, and only inadequately with landscapes. Normally attributable to known dates and authors, the cathedral, the painting, the silver goblet possess a type of specificity lacking in the lake, the meadow, or the mountain range. Relics of antiquity are mentally segregated even when seen in their locales; few plants, animals, or landscape features are segregated in the same fashion, and mobility or seasonal change renders them still more amorphous. Not even a landscape's boundaries can be definitely specified; they shift with every perspective, the landscape merging indissolubly into its surroundings.[3]

Differing attachments to nature and to culture

Such differences affect modes of attachment to each realm. On the whole, we admire nature as previous to history and at the same time one with the present. Wild nature especially offers at once the age-old and the immediate, its configurations and seasonal and diurnal rhythms comfortingly familiar, yet also awesomely primordial. By contrast, the fascination of historical vestiges lies in the unique confrontations, the unanticipated contingencies, and the transience of things past.

Natural and cultural heritage also elicit differing modes of communion. Nature seems essentially *other* than us: we may feel at one with its life-supporting fabric, but we seldom put ourselves in nature's place or project ourselves into non-human lives. By contrast, the cultural heritage promotes empathetic communion. Our very lineaments augment the legacy of our progenitors. The genealogical specificity of our own ancestors imbues the human heritage with personal allure. However deeply we may love nature, most of us identify more easily with human relics and, when necessary, rise more readily to their defence.

Hence crusades to save antiquities are more likely to enlist local support than those for the sake of nature. It was outsiders who spearheaded the campaign to preserve Tasmania's southwestern wilderness and prehistoric sites, while the Tasmanian government would have scuttled it in the interest of hydroelectric development and local employment.[4] It is the international community of ecologists and biologists who worry about development programmes that threaten the Amazon rain forest, its largely untapped gene pool, and global evapotranspiration balances; whereas Brazilians focus mainly on the profits in mining, agriculture, and industry. (Yet local unconcern also imperilled Pharaonic antiquities in nineteenth-century Egypt: only export to the British Museum saved mummies from being sold off for medicinal use.[5] And though every nation today trumpets devotion to its precious cultural heritage and seeks its restitution from foreign plunderers or purchasers, entrepreneurial greed and local need often subvert cultural nationalism with virtual impunity.)

Unlike much of the cultural heritage, most of the natural heritage cannot be exported; its value inheres almost wholly in its locale. Indeed, it does not even enter the realm of calculable commodities – a realm that increasingly dominates awareness of artifactual relics. Almost the first point mentioned when a cultural relic comes on the market is the record-breaking sale price. If collectors could buy ecosystems as they do paintings and bronzes, they might value the natural heritage more highly; but having to leave the ecosystem undisturbed *in situ* would deter all but the most selfless philanthropists.

Differing rationales for preservation

Differently perceived and evaluated, nature and culture are also safeguarded for different reasons. Most arguments advanced for nature conservation dwell on long-term economic or ecological benefits; most arguments for antiquities preservation cite cultural or aesthetic benefits. Exceptions abound, to be sure: historic preservation is often justified on the grounds that it saves energy and materials and generates tourist revenues, but aesthetic, emotional, and spiritual rewards are widely felt to carry more conviction than pragmatic considerations. As for nature conservation, the bird and wildlife lobbies are uniquely vehement. And the US National Parks system came into being primarily for inspirational and recreational aims that still buttress arguments against their industrial or mineral exploitation or residential development. But only a minority speak of saving nature on religious or aesthetic or ethical grounds. Conservation spokesmen focus instead on the importance of environments and ecologies as life-support systems whose loss would cripple agricultural and medical progress and ultimately imperil humanity. In short, they rely on compelling fears about natural resources and balances. Such fears play no appreciable role in campaigns for historic preservation.

III   ORIGINS OF CONSERVATION AND PRESERVATION

Before nature and antiquity could be treasured, they had first to be recognised as realms distinct from, but as real as, the everyday present. That revolution in thought had its origins in the Renaissance. To the medieval mind, the manifestations of external nature and the vestiges of the past had been little more than reflections of their own mundane world, or unwelcome distractions from the world of the spirit. It is no accident that the glories of nature and of antiquity both found their first post-classical celebrant in Petrarch, the first great humanist. Petrarch's emotional involvements, with the Alps and with those great peaks of the human mind, the writings of classical antiquity, were alike at odds with his Christian faith – a conflict he long struggled to resolve.[6]

New attitudes toward nature and antiquity

From the fifteenth century on, manifestations of nature and of antiquity more and more absorbed European attention. At home, the extension of civic authority and domestic order made travel less perilous and more pleasurable; abroad, exploration continually opened up new marvels for study and experience. Yet the fate of relics of either realm still aroused little concern. Manuscripts apart, humanists seldom preserved fragments of material antiquity as such; rather they restored, amalgamated, and emulated them for use in modern structures. Convinced that decay permeated all the earth's features and creatures, seventeenth-century savants never contemplated intervening to stem the supposedly ineluctable process of secular dissolution.[7]

By the late eighteenth century geographers and archaeologists had revealed a congeries of exotic wonders, past and present. But these new worlds of nature and of antiquity elicited quite different responses. In the realm of nature, the visible benefits of human intervention, notably in agriculture and engineering, led men to celebrate their impact on the face of the earth and to welcome even more far-reaching environmental transformations. The *philosophes* viewed untended nature as hideous and wasteful: God had deliberately left Creation unfinished so that human skills would perfect it. In the realm of antiquity, however, finds from Pompeii and Herculaneum to Egypt and the Aegean revealed fascinating but generally benighted pasts. The remnants of those pasts were exotic curiosities rather than precious relics of ancient genius.

Not until the nineteenth century did the pace of change and Romantic sentiment render both nature and antiquity sacred and promote impulses to protect their vestiges against not only decay and dissolution but improvidence and iconoclasm. But concern for nature and for relics of the past diverged in rationale and in timing.

The birth of historic preservation

Sentiment for preserving antiquities arose out of wholesale transformations of society and thought in the late eighteenth and early nineteenth centuries. The French Revolution and the Napoleonic Wars made even recent times seem remote and irrecoverable; the Industrial Revolution distanced them still further. The rise of nationalist sentiment fostered attachment to ancient monuments as symbols of collective identity; antiquities gained credence and appeal as historical witnesses more reliable and more compelling than documents; the essays of Rousseau and the poems of Wordsworth brought to widespread consciousness the physical souvenirs along with the scenic locales of childhood. All these developments helped induce a climate that favoured venerating and preserving ancient buildings and monumental remains.[8]

The birth of ecological concern

Nature conservation emerged out of different circumstances a half century later. European engineers, surveyors, and naturalists during the eighteenth century had on occasion observed that deforestation and damming induced erosion, flooding, and avalanches.[9] Not until the 1840s, however, viewing the dwindling forest cover and turbulent river regimes of his native Vermont, did the American scholar George Perkins Marsh recognise the generally deleterious consequences of human impact on the environment and suggest broad measures to ameliorate them.[10] In his *Man and Nature* (1864) Marsh extended these conclusions. Tracing environmental degradation from the Roman Empire to his own day, from the deserts of North Africa to the torrents of the Alps, he warned that without drastic reform mankind would reduce the habitable globe to the barrenness of the moon.[11]

Man and Nature pioneered ecological awareness over the next half century. Environmental impact came to be seen in a new light; man the improver turned into man the destroyer who unwittingly subverted the balance of nature; only ecological stewardship would prevent the technology that had subdued the earth from wrecking the organic bases of civilisation.

The impact of nineteenth-century industrial development

Pressures to preserve the past and to conserve natural resources gained urgency and began to converge as industrialisation and urbanisation threatened to obliterate both the natural environment and the remnants of antiquity. Nostalgia for what was seen as ancient and stable focussed on pre-industrial life and landscapes. Men cooped up in cities and factories were deprived of a birthright, part classical and Arcadian, part wild or rustic, whose surviving vestiges might yet restore social health if carefully husbanded.

Concern for protecting nature and antiquities peaked around the turn of the century in both Europe and the United States, but transatlantic emphases differed. Americans, who had exalted nature in place of history even while gutting the wilderness, evinced keenest concern for these ravaged lands and for still unspoiled tracts of wilderness. Their conservation efforts were of two main kinds: afforestation to restore the balance of vegetative cover, to equalise run-off, and to reduce erosion; and the establishment of national parks as enduring oases of recreation and refreshment remote from the turmoil of the busy man-made world. The latter involved nature preservation on an unprecedented scale.[12]

By contrast, Europeans long bereft of pristine environments turned for solace to the man-made landscape of pre-industrial times, treasuring its remnant features and folkways and replicating their forms. The last decades of the century saw the birth in Britain of the Society for the Protection of Ancient Buildings and the first Ancient Monuments Act to safeguard notable antiquities; France and Germany spurred comparable measures to preserve the tangible past.[13]

Yet each continent mounted protective action of both kinds.

Americans in the 1880s and 1890s revived Colonial architecture and refurbished historic houses for display. Britain in the same period promoted countryside concerns through the National Trust and *Country Life*. Everywhere the conservation of nature went hand in hand with the preservation of antiquities.

Moreover the same eminent figures often played prominent roles in both crusades. In eighteenth-century France, Volney celebrated ruins and nature alike; in the nineteenth century Chateaubriand and Hugo led movements to preserve both realms from depredation and neglect.[14] In nineteenth-century Britain Wordsworth, Ruskin, and Morris evinced parallel concerns for nature and for the past. In the United States Marsh, who pioneered nature conservation, also urged Americans to save and display relics of husbandry and industry and domestic life as evocative reminders of how their forbears had forged a new nation.[15]

IV   STEWARDSHIP AND SELF-INTEREST

Similar ideals of stewardship animated nineteenth-century crusades to protect both aspects of the patrimony. Nature and antiquities alike were seen as inheritances not to be consumed but kept in trust for generations to come. Ruskin's *Modern Painters* and Marsh's *Man and Nature* contain strikingly parallel phrases: 'Old buildings are not ours. They belong partly to those who built them, and partly to all the generations of mankind who are to follow us', wrote Ruskin.[16] And Marsh: 'Man has too long forgotten that the earth was given to him for usufruct alone, not for consumption, still less for profligate waste.'[17] The notion of heritage as an enduring legacy from every past to every future was uppermost in Marsh's mind. America's rich resources came by covenant from a divine creator who had willed a chosen people an unblemished continent, then handed on by Founding Fathers who had secured that heritage for their descendants. The legacy must be protected and restored not only out of self-interest but as a sacred duty.

Nineteenth-century crusaders felt obligated to previous and to future generations partly because they cherished their own posthumous reputations. To be well regarded by posterity was a recurrent *leitmotif*. Today it is seldom evoked. And with it has gone a major prop for the idea of stewardship. The harbinger of that loss was the American senator at the beginning of the century who was asked to help reserve a large tract of National Park land for posterity, and responded, 'What's posterity ever done for me?' The modern attitude toward posterity is exemplified in a cartoon showing the dismay of expectant inheritors as a lawyer begins to read the deceased's will: 'Being of sound mind and body, I blew it all.' Prospects of global annihilation or instant Armageddon have perhaps curtailed commitments to a future which may never come. At all events, immediate or imminent self-interest now dominates many, perhaps most, appeals for conservation and preservation.

Reliance on self-interest seems to me a tactical if not an ethical error. Emphasis on resource and energy savings and tax-break advantages leaves the American historic preservation movement at the mercy of market forces that may at any time wipe out the benefits that currently accrue to preservation. Servicing the aims of development speculators sacrifices a coherent preservation rationale to immediate pay-offs.[18] It should be remembered that preservation's initiating impulse is not short-term economic gain but the fundamental enhancement of life in the longer run.

Nature conservation relies still more on enlightened self-interest. Whether they seek to curtail pollution, to maintain diverse gene pools, or to prevent the extinction of species, most conservation advocates couch their arguments in hard-headed pragmatic terms based on scientific expertise. But such arguments can engage the minds or sway the hearts of relatively few people anywhere. It might be more effective to emphasise the attachments to nature we all share through genetic inheritance, personal experience, and beloved evocations in poetry, prose, and painting. Like traces of the past, vestiges of nature immeasurably enhance life. To be prized, nature need not be wild, awesome, rare, or beautiful; our lives would be the poorer without even its most prosaic and familiar forms.

V AWARENESS OF INTERFERENCE

Arguments for historic preservation transcend self-interest more readily than those for nature conservation. But nature conservators some time ago achieved an insight that still eludes most guardians of historic relics – the awareness that human impact, their own included, profoundly and inevitably alters the objects of their concern. It had been the accepted view that the earth made man; but as Marsh and others came to see, man also made the earth.[19] Intentionally or otherwise, human agencies continually and often irrevocably change the face of the globe. It is now common knowledge that we can return neither to a state of nature nor to any supposed 'balance'; that environmental interference always requires further interference; and that stewardship of the natural heritage means not leaving nature to her own devices but interfering more carefully. Indeed, we could not leave nature to her own devices even if we wished to, for manipulating the environment is integral both to the nature of man and to the condition of nature. Recognising the inevitability of such interference was a necessary first step toward rectifying previous errors and shaping future interference along lines more benign, constructive, and, where possible, reversible.

Few advocates of historic preservation yet realise that just as we reshape nature we also reshape our past. We can no more avoid interfering with the relics of antiquity than with the world of nature, however benign our neglect or well-intentioned our intervention. But most preservationists like to believe that they are saving the 'real' past, whose authenticity they treasure. In their view, the preserved object ought to be as it originally was, aside from natural erosion and accident. But every survival in fact attests not only its original state but all subsequent alterations and additions, deliberate or unintended. Nothing is just as it was, for time has revised even our perceptions of it. We use such remnants most fruitfully when we realise that they are continually being reformed; indeed, our very efforts to preserve them attest their malleability. To be sure, what our predecessors have left us should be respected, but a legacy simply preserved is a dead weight; we must accept that we also continually domesticate the past. Like modern nature, antiquity is in large part an artifact of the present.

Seeking to repair the injuries wreaked on the earth's surface, Marsh argued that man was a free moral agent, acting above nature.[20] I would similarly argue that although formed by the past we are not wholly bound by it; as it has made us, so we go on remaking it. Respecting its messages by revising their meanings, we best appreciate the riches of the past by ceaselessly striving to loosen its bonds.

Notes

1. Founded in 1967 in Barbados, still its headquarters, the Association publishes the monthly *Caribbean Conservation News*.
2. Lord Charteris of Amisfield, 'The Work of the National Heritage Memorial Fund', *Journal of the Royal Society of Arts*, 132 (1984) pp. 325–38.
3. David Lowenthal, 'Finding Valued Landscapes', *Progress in Human Geography*, 2 (1978) pp. 375–418.
4. K. B. Ryan, 'Constitutional Law and the Preservation of Wilderness', meeting of the Association of American Geographers, Washington, DC, 1984. See also, S. Robert Aiken and Colin H. Leigh, 'Hydro-electric Power and Wilderness Protection', *Impact of Science on Society*, 36 (1986) pp. 85–96.
5. Brian M. Fagan, *The Rape of the Nile* (London: Macdonald & Jane's, 1977) pp. 44–7.
6. Angelo Mazzocco, 'The Antiquarianism of Francesco Petrarca', *Journal of Medieval and Renaissance Studies*, 7 (1977), pp. 203–24.
7. David Lowenthal, *The Past is a Foreign Country* (Cambridge: Cambridge University Press, 1985), pp. 75–88, 390–1.
8. Ibid., pp. 391–5.
9. David Lowenthal, 'Introduction', in George Perkins Marsh, *Man and Nature* (1864) (Cambridge, Mass.: Harvard University Press, 1965), pp. xviii–xix.
10. George P. Marsh, *Address Delivered before the Agricultural Society of Rutland County*, September 30, 1847 (Rutland, Vt., 1848).
11. Marsh, *Man and Nature* (1864).
12. Roderick Nash, *Wilderness and the American Mind* (New Haven: Yale University Press, 1967) pp. 108–60; Peter J. Schmitt, *Back to Nature: The Arcadian Myth in Urban America* (New York: Oxford University Press, 1969).
13. Martin J. Wiener, *English Culture and the Decline of the Industrial Spirit, 1850–1980* (Cambridge: Cambridge University Press, 1981) pp. 44–70; Jan Marsh, *Back to the Land: The Pastoral Impulse in Victorian England* (London: Quartet, 1982).
14. Pierre de Lagarde, *La Mémoire des le Pierres* (Paris: Albin Michel, 1979) pp. 54–78.
15. George P. Marsh, *The American Historical School* (Troy, New York, 1847); David Lowenthal, *George Perkins Marsh* (New York: Columbia University Press, 1958) pp. 101–3.
16. John Ruskin, *The Seven Lamps of Architecture* (1849), ch. 6, sec. 20 (New York: Farrar, Straus and Cudahy/Noonday Press, 1961) p. 186.
17. Marsh, *Man and Nature*, p. 36.
18. Richard W. Longstreth, 'Preservation's Exposed Flank', *Historic Preservation* 32:6 (1980) pp. 54–5.
19. *Man and Nature* was written to show that 'whereas Ritter and Guyot think that the earth made man, man in fact made the earth' (Marsh to Spencer F. Baird, 6 March 1860, Spencer F. Baird Correspondence,

Smithsonian Institution, Washington, D.C.) See Lowenthal, *George Perkins Marsh*, p. 248.

20. Lowenthal, 'Introduction', in Marsh, *Man and Nature*, pp. xxiii–xxiv; George Perkins Marsh, 'The Study of Nature', *Christian Examiner*, 68 (1860), p. 34.

8 Epilogue: Reflections on the Role of Ideological Perceptions

Erik Baark and Uno Svedin

Many valuable approaches to the study of man, nature and technology have been presented in the foregoing essays contributed by scholars from diverse disciplinary backgrounds. In this epilogue, we shall therefore limit ourselves to some concluding reflections related to the role of ideological perceptions.

I CAUSAL LINKAGES?

Concepts of nature, culture and technology have always been elusive. Most of the contributors to this volume have therefore resisted the temptation to define these concepts. In fact, a large variety of meanings associated with each of these terms has appeared in the contributions. Instead, the search for a better understanding has been oriented towards linkages, or relationships, between man, nature and technology.

In many respects, the study of linkages provides an interesting means of grasping the inherent characteristics of nature and technology. One important question is, however, whether these relationships are causal. As the survey conducted by Kates demonstrates, causal theories have a long history. These theories cover a wide range of ideas: on the one hand, rather simplistic, one-dimensional theories of, for instance, supernatural control have been very popular; on the other hand, a fundamental causality is proposed in partial theories such as the famous Malthusian forecast of a diminishing per capita availability of natural resources such as food, due to population growth.

The idea of causality has been particularly prominent in discussions of materialist versus idealist theories. An example of a materialist theory is the theory of hydraulic societies which is examined in Worster's essay. Here it is argued that social organisation and power

135

relations are, to a large extent, determined by material conditions. The study indicates the way in which the mastery of adverse natural conditions in arid zones has fostered more elitist and centralised patterns of social relations.

Another argument which reflects a causal relationship of a materialist outlook is that technology embodies the roots of certain patterns of social organisations. This notion, which is popularly known as 'technological determinism', is discussed by Wynne in his essay on technology as cultural process. His conclusion is, however, that although 'technology does have intrinsic force and that this may well encompass and freeze, in its own way, the whole field of possibility for some societies or groups receiving a technology, the cultural process model does not commit the often-ensuing slide into technological determinism as a model of history'.

While visualising the linkages between nature, culture and technology in the light of causal processes can be rewarding in many instances, one should also be keenly aware of the limitations involved. The preoccupation with the identification of causal factors frequently represents a methodological bias inherited from the natural sciences. Interpretations of the multi-causal type have been more acceptable in the academic disciplines of humanities and social sciences. Modern systems science has also been important for the revival of research methodologies which emphasise the multiplicity and interdependence of factors.

The difficulties of defending a unique causal relationship between man, nature and technology can be partly relieved by accepting the normative nature of dominant concepts. Thus Kates concludes that objective criteria for the assessment of a theory hardly exist, and instead proposes the search for a theory which is designed to serve normative purposes, that is, a prescriptive theory. He suggests a theory which could outline, for example, the transition from the globally unbalanced industrial stage of today to a desirable steady-state population situation by the next century.

This is the reason for choosing the theme of ideological perceptions as a key to understanding the normative nature of the various conceptions of man's relationship to nature and technology. We find, for instance, that concepts of rationality and causality remain contingent to the ideological position of the observer. Economic rationality is a typical example of a concept which may be used in ideologically biased ways, for instance in connection with decisions concerning the use of natural resources, the environment, or technological change.

This sort of 'rationality' is, as Wynne points out in his essay, mostly defined by public organisations. Alternative interpretations – for example, those articulated by people affected by changes, but unable to formulate themselves in 'expert' terms – are ignored. But nevertheless such alternative concepts are valid, and indeed, valuable for decision-making in the face of uncertainty and great risks.

II POWER

It is conceivable that one could discuss ideologies *per se* without any particular reference to aspects of power. However, when the role of ideologies in relation to Man's perception of nature and technology is examined, it becomes apparent that power relations constitute a key issue. The distribution of power in society often provides for platforms of ideological persuasion and the means for political acts related to Man's environment.

A good example is the idea of a skill-controlled cornucopia, which Thompson has associated with entrepreneurs as a social group. This perception of nature can be regarded as an expression of ideologies designed to maintain the growth of powerful organisations engaged in the exploitation of natural resources. The multinational oil companies, for instance, have typically based decisions for investment in prospecting activities on the widespread belief in growth appearing in times of economic boom, as, for example, during the 1950s and 1960s. Such activities have often led to a dependence on the oil companies for economic development, a position which has tended to strengthen their power in both oil-producing and oil-consuming societies.

However, the perception of scarcity has been equally important as a foundation for centres of power and, interestingly, the oil industry may be utilised as an example in this respect too. A basis for the establishment of a cartel between oil-producing countries has, in a sense, been the notion of scarcity of oil resources which was associated with a number of oil crises during the 1970s. An ideology of administering scarce resources which developed during this period appears to have been essential to the maintenance of common interests in the cartel. When perceptions slowly reverted in the early

1980s to the original ideas of cornucopia, and when the vision of immediate scarcity waned as new oil-producing areas flourished, the power of the oil-producing cartel of the 1970s was quickly eroded.

Ideologies are generated not only by powerful elites, but also by those in opposition to them. Defence of the local environment and support for technologies that people can understand and control is often expressed through myths and imagery – the technological animism that Wynne discusses in his essay. According to Wynne such imagery serves to provide security by means of a reduction of technological systems to 'creatures'. The technologies of the nuclear power plant, the computer system and the great dam are all unmanageable, when seen from the perspective of ordinary citizens. Their reaction is often to try to deal with the threats from such technologies by regarding them as supernatural beings. This imagery may thus constitute the key element of a 'defensive ideology'.

Lowenthal examines a different, but crucial aspect of power relations in the sphere of interrelationships between Man, nature and technology. Frequently centralised power will go hand in hand with ideologies of exploitation of nature. However, the capacity to protect the natural or man-made environment may also require centralised power in order to counteract the prevailing market forces. Under such circumstances the role of state authorities and the ideology which legitimises this centralised power may be crucial for maintaining the environment. The situation may thus encourage the central authorities to form an alliance with marginalised segments of the population against the forces of 'economic progress'.

The ideas of control of nature and technology which give rise to such power relations frequently involve particular perceptions of risks. The risks of natural disaster or technological havoc may be played down by the powerful elites who base their position on rapid development. For their opponents, the ideology of development entailing a calculated probabilistic risk may be entirely rejected. Another reaction may be to ignore the risks and their consequences simply because individuals perceive themselves as totally without influence.

III   THE UTOPIAN DIMENSION

In many of the essays in this book we find a direct or an indirect reference to Utopias; these references will frequently be made as an indirect advocacy for a specific type of society, the Utopian vision being utilised to bring out this dimension.

Worster discusses the Utopian 'vision of a future beyond domination' which was proposed by Herbert Marcuse. This was the idea of replacing the domination of capitalist industrial society with rational gratification, a project which would lead to the destruction of power elites and a democratisation of self-fulfilment. Worster remains sympathetic to the Utopian vision of relieving people of domination; but at the same time, he criticises Marcuse for failing to realise that the provision of even the limited resources required for 'rational gratification' implies the continued domination of the environment.

The normative perspective of Utopian visions is sometimes hidden in the framework of a neutral, objective analysis, but generally the contributors have formulated their partisanship in an explicit manner. The essay by Sagasti, for example, combines Utopian ideals with the ambition of creating a new, more realistic model of the role of science and technology in development. A major point in Sagasti's line of reasoning is the need to recognise the potential importance of non-Western modes of development. The alternative to utilising the Western civilisation as a frame of reference for evaluating development elsewhere in the world is, Sagasti argues, to start a process of emancipation which is directed towards the creation of an endogenous scientific and technological base. This process is seen as contingent to the positive interaction between three currents of human activities (evolution of speculative thought, transformation of the technological base, and modification of productive and service activities). On the basis of this conceptual framework, Sagasti maintains, the present situation of backwardness of developing countries can be explained; in addition, the framework should make it possible to design strategies for overcoming backwardness.

The Utopian ideal underlying Sagasti's presentation is thus an emancipated South. This concept of the South is perceived as the situation in which development processes in countries of the non-Western world are generated and maintained independently of the countries in the North.

In both Thompson's and Wynne's contributions the issue of nuclear power generation is utilised as an important example. Non-military nuclear power generation is certainly a fact of life for many people in the advanced industrialised countries. Nevertheless the images which this particular technology conveys are linked with both positive dreams of the future, that is, Utopias, and visions with strong negative connotations, that is, dystopias.

In the 1950s the prospect of the non-military use of nuclear energy came to be portrayed in terms of a Utopia based on an immense abundance of energy. The technology was presented to the public as the key solution to the world's long-term energy requirements. Twenty years later, the image had undergone a major transformation, as the environmental hazards of defective nuclear plants and the disposal of nuclear waste aroused the scepticism – if not outright hostility – of large sections of the populace against the technology. Nuclear power thus conjures up alternative images – Utopias or dystopias – for different groups in society. Simultaneously, these images may be projections of desirable futures; they become concrete expressions of cultures, of ideological interpretations of the relationships between nature, culture and technology.

Nuclear power is a good example of linkages and their cultural interpretations. It can be seen merely as a highly efficient technology for the extraction of energy for our society, utilising to the full Man's mental capabilities, his ability to understand nature in a scientific manner and to design techniques for manipulating and controlling nature's resources. Indeed, the belief in the ability to use abstract knowledge for practical pursuits is a distinctive feature of modern Western civilisation. In many other cultures people appear to have been more restrained in their faith in man's ability to understand and control nature.

In contrast to the modern confidence in man and his power over nature and technology, epitomised in the development of nuclear power stations, we find the critique of modern insensitivity to the exploitation of nature, for example, the strip mining of large areas for uranium. Moreover, it can be regarded as a mirror image of social relations, that is, the hierarchical and centralised modes of decision-making. The essential point is not the use of particular technologies and the attitudes to natural resources as such, but the hopes or fears which are reflected in the images surrounding these. Images of Utopias (and dystopias) frequently manifest the qualities associated with interrelationships between nature, culture and technology. It is always difficult fully to understand contemporary society and culture. We would argue, however, that through the scrutiny of the images of the Utopias and dystopias of our times, we can acquire a more comprehensive view of the cultural dimensions of man's relationship to nature and technology.

Index